P9-AFT-379

Fitch, John E.
Marine food and game
fishes of California /
1971.
33305202966531
CU   06/02/09

28

CALIFORNIA
MARINE FOOD
AND GAME FISHES

BY

JOHN E. FITCH

California Department of Fish and Game

AND

ROBERT J. LAVENBERG

Los Angeles County Museum of Natural History

Illustrations by Evie Templeton

Photographs by Daniel W. Gotshall and Charles H. Turner

459771

UNIVERSITY OF CALIFORNIA PRESS

BERKELEY, LOS ANGELES, LONDON   1971

SANTA CLARA COUNTY LIBRARY

3 3305 20296 6531

SANTA CLARA COUNTY LIBRARY
SAN JOSE, CALIFORNIA

Dedicated to CHARLES H. TURNER
May 15, 1934 to October 27, 1970

UNIVERSITY OF CALIFORNIA PRESS
BERKELEY AND LOS ANGELES, CALIFORNIA

UNIVERSITY OF CALIFORNIA PRESS, LTD.
LONDON, ENGLAND

ISBN: 0-520-01831-1
LIBRARY OF CONGRESS CATALOG CARD NO.: 73-132419

©1971 BY THE REGENTS OF THE UNIVERSITY OF CALIFORNIA
PRINTED IN THE UNITED STATES OF AMERICA

2 3 4 5 6 7 8 9

CONTENTS

Illustration on Cover: Black-and-yellow rockcod

Coos Bay, Oregon

Eureka

San Francisco

Santa Cruz

Morro Bay

Diablo Cove

Pt. Conception

Santa Barbara

Ventura

Palos Verdes
San Pedro
Newport Beach

San Diego

Ensenada

N

[4]

INTRODUCTION

California's marine fishes have been important to man's welfare ever since some enterprising early inhabitant found that there was food to be gotten from the exposed shoreline, when the tide was out, to supplement his meager land-based diet. The first inhabitants of our coast were Indians, and according to radiocarbon dating techniques, they arrived here about 9,000 years ago.

Microscopic examination of screened residue from the Indians' campsites and garbage dumps (middens) has revealed much information about how they lived, what they ate, and the tools they used to obtain their food. A Chumash village site at Ventura yielded the remains of forty-five species of fishes, including thirteen kinds of sharks, skates, and stingrays. Investigators have speculated that the Indians used hooks, traps, gill nets, harpoons, and beach seines, besides their hands, to catch these fish, and that they must have had far-ranging, fast-moving craft to reach some of the fishing grounds.

North of Point Conception an especially interesting undisturbed midden was discovered above a rocky cliff. Residue from a portion of it was examined by 4-inch increments to a depth of 10 feet. The bottommost 4-inch layer, representing the earliest years of occupation, contained remains of fishes that live in the intertidal area, and it seems probable that the Indians had caught these by hand while scrounging for food among the rocks at low tide. While searching along the shoreline the Indians must have observed other kinds of fishes swimming just out of reach and have wanted to harvest some,

[5]

which undoubtedly led to the fabrication of harpoons or spears. Otoliths (ear stones) and other recognizable remains from these harder-to-obtain fishes were found a foot or more up in the midden, providing evidence of an improved technology.

The quantity and quality of fish remains that turned up at progressively higher levels in the midden indicate that as the Indians became increasingly familiar with the sea and its bounteous supply of fishes, their fishing gear and techniques must have improved even more. There is evidence that the Indians used mobile craft to fish the nearshore kelp beds. The find of an abundance of night smelt otoliths is a sure sign the tribesmen had discovered spawning runs of these tasty fishes on nearby sandy beaches. Nowhere in this midden, however, was there an indication that the inhabitants had ventured much beyond the outermost edge of the fringing kelp. Apparently the seas between Point Arguello and Morro Bay were just as inhospitable during the 9,000 years the Indians occupied the area as they are today.

In the twentieth century, fishing in the coastal waters of California continues to be an important part of man's activities, but today there is great emphasis on the sporting aspects. We feel certain that edibility was the only criterion the aboriginal Indians considered in categorizing their daily fish harvest. The tastiness of a fish is still of prime concern, but the modern sport fisherman gives extra status to a species that ranks high in elusiveness, fighting ability when hooked, and "award" or "reward" potential. When sport and commercial interests are competing strongly for a given species, and research shows there are not enough of that species to supply the needs of both, regulations often have been enacted to reserve the species in question for the exclusive use of the recreational fisherman. Thus, over the past fifty-odd years, sport fishermen in California have been granted the sole right to harvest striped marlin, spotfin croakers, California corbina, kelp bass, striped bass, and several others.

[6]

We have no way to measure the Indians' annual fish harvest, but since 1910 catch statistics have been maintained for California's commercial fisheries and, since 1946, for sport partyboat fishermen. In addition, surveys of other marine sport fisheries (e.g., shore, pier, skiff, private boat, and skindiving) have been conducted on several occasions during the past decade or two. The sport fisherman's take is reported in numbers, but commercial landings are given in pounds, and live bait fishermen use "scoops" in noting their catch, so it is impossible to come up with much more than an educated guess as to the total number of fish removed by man from the ocean off California each year.

During recent years, sport fishermen have hooked over 7 million fish annually in southern California, and more than 3 million more north of Point Conception. In addition, sport fishermen net about 750,000 pounds of surf smelt and night smelt per year, and this poundage converted into numbers swells the sport angler's bag by about 10 million fish. Possibly one hundred fifty species are involved in the marine sport catch, but only about thirty kinds are truly important, and fewer than ten of these often contribute upwards of half the total poundage.

Between 10 and 15 million pounds of fish (mostly anchovies) are used for live bait each year in the ocean. At an average of twenty per pound, the live bait fishery removes a minimum of a quarter billion fish per year for use by sport fishermen.

Because cannery landings (i.e., anchovies, bonito, mackerels, sardines, and tunas) fluctuate widely from year to year depending upon availability as well as demand, and because some fish (anchovies) will weigh but a fraction of an ounce each while others (tuna) will average 10 to 30 pounds each, it would be impossible to calculate an annual average for the commercial catch that would be reliable. Perhaps it will be sufficient to point out that during a single fishing season (1936-1937)

[7]

more than 4 billion sardines (791,000 tons) were estimated to have been caught by California's commercial fishermen. Tops for albacore during a single year (1950) would be about 4½ million fish. When we add to this some 20 to 40 million mackerel, 10 million flatfish, 5 million bonito, 3 million rockcod, and a few million for salmon, sablefish, hake, white seabass, and a wide assortment of other marketed fishes, one ends up with a rather staggering total.

Among the families we are covering in this book, anchovies are harvested in the greatest numbers each year, whereas salmon and albacore are probably the most sought after or desirable. Swordfish attain a greater length (sword included) than any of the other food and game fishes, but they run second to molas for weight. Giant sea bass, marlin, bigeye tuna, louvars, and bluefin tuna are middleweights and lightweights by comparison to a record-sized mola, but to date, giant molas have avoided our coastal waters. The International Game Fish Association, Fort Lauderdale, Florida, produces each year a listing of recognized world-record game fishes including the names of the individuals who caught the various species and the locality of capture. California has produced a fair share of these records.

Insecticide residues, lead compounds, industrial and domestic wastes, oil, and other pollutants constantly pour or rain into the world oceans, so the fishes and other creatures inhabiting our waters may not survive long enough to reach record sizes in future years. In 1969, for the first time in history and less than forty years after mass production and use of DDT and other potent "bug" sprays a tolerance level was established as to the amount of insecticide residues that would be permitted in fish intended for human consumption. These residues, which result from crop-dusting and spraying operations on land, are carried into the sea primarily by air currents and by runoff from irrigation and rainfall. In the sea, these residues are taken in by planktonic forms

[8]

at the lowest end of the food chain and transferred from predator to predator until eventually they are ingested by man at the top of the food chain. The residues are not utilized for food by any of the organisms, but they are stored and concentrated in fat and other body tissues. Thus, when man eats a tuna, he is getting the "benefit" of insecticide residues picked up and stored by thousands of prey species which had been eaten by the tuna, which in turn had eaten tens of thousands of smaller "contaminated" prey, and so on down the line to the planktonic forms. Because these residues become more concentrated with each higher link in the food chain, many of the large food and game fishes in our oceans already contain enough of these toxic compounds in their flesh to alarm public health officials and medical authorities. Unfortunately no one is capable of making (or willing to make) a positive statement as to what constitutes an unsafe level for man, or whether the end result will be a "lingering death," "sterility," "gene mutations in future generations," or a continued "normal" existence of paying taxes and fighting wars.

SCOPE OF COVERAGE

We have attempted to include all families of marine teleost fishes in which one or more species are important for their edibility, for the sport of catching them, or for both reasons. In some large families very few members can be considered "food" or "game" fishes, but we have included all species in our checklist if they are strictly marine. Thus in the family Cottidae, the cabezon is perhaps the only truly important food or game fish, yet we have listed more than forty species of sculpins, leaving out only those which belong to the freshwater genus *Cottus*.

We have not included any of the saltwater fishes that are restricted to Imperial Valley's irrigation canals and Salton Sea (e.g., *Elops affinis, Bairdiella icistia, Cyno-*

[9]

scion xanthulus), nor have we mentioned any of several introduced species that failed to survive in our coastal waters.

We might be taken to task for including the remora, the prowfish, and a few others, but we believe they fit better into this book than into our previous volume, *Deep-water Teleostean Fishes of California* (California Natural History Guides: 25, 1968) or a subsequent book on miscellaneous small fishes. Members of several families that were covered in *Deep-water Fishes* make excellent table fare or are noted for their fighting abilities when hooked, so some overlap has been unavoidable. Since the pomfret and cutlassfish families (Bramidae and Trichiuridae) contain species other than those we previously discussed and illustrated, we were able to avoid duplication in these two cases, but were forced to repeat with the opah, louvar, and ragfish — sole members of their respective families.

We have illustrated one or two of the more important members of each family, and have attempted to explain how the figured species can be distinguished from other fishes. Our natural history notes are brief, yet they are as complete as available information permitted us to make them. In many instances, we found it necessary to examine stomach contents, check scales or otoliths for age, inspect gonads for signs of maturity, and measure and weigh large individuals in order to present factual information where little or none was available. Whenever possible, we have included catch data for both sport and commercial fisheries, and we usually have mentioned the kind of gear that is most successful for catching the species discussed. If there are fewer than eight to ten species in a particular family, we have constructed a rather simple one- or two-character key for distinguishing them. Since identification keys are never infallible (e.g., there are always a few oddballs that will not fit; a new or previously unreported species occasionally is caught; etc.), other authorities should be checked

[10]

for puzzlers. For most of the larger families (i.e., Pleuro-nectidae, Carangidae, Scombridae, Embiotocidae, Scorpaenidae, and Cottidae), technical publications are available, although not always up to date, to aid in iden-tification. We have included references to these in our list of suggested reading.

All the drawings are based either upon freshly pre-served specimens or, for large unwieldy fishes, on photo-graphs of freshly caught individuals. Various body pro-portions, fin lengths, fin positions, and other anatomical structures are depicted as they appeared on the speci-men drawn, and these are believed to be typical for the species.

We have presented the family accounts alphabeti-cally (Albulidae through Zaproridae) in order to sim-plify the task of finding a family without having to look in the index first.

As was the case with *Deep-water Teleostean Fishes of California*, we received helpful suggestions from many quarters, but the ultimate choice was ours as to which species to include and which to omit.

A BRIEF HISTORY OF ICHTHYOLOGY AS IT
CONCERNS
CALIFORNIA'S FOOD AND GAME FISHES

Although North American marine ichthyology dates to 1814, at about the founding of the Philadelphia Acad-emy of Sciences, when Samuel L. Mitchell (1764-1831) published a small tract on the fishes of New York, none of the California fish fauna had been discovered, named, or described prior to 1850. California's ichthyology dates to 1851, when a young doctor, William O. Ayres (1817-1891), joined in the gold rush to California, and became one of the first physicians in San Francisco and a founding member of the California Academy of Sci-ences. Ayres, a Boston medical colleague of David H. Storer, gathered fishes from the San Francisco markets and California coast and pioneered in discovering and describing food and game species. Twenty-eight fishes named by Ayres are still recognized today, including such diverse species as the longfin smelt, grunion, white seabass, and giant sea bass.

There were three major surveys in the United States during the mid-nineteenth century that made collections of the local flora and fauna: the Pacific Railroad Survey, Exploration of the Mexican Boundary, and Exploration of the Western Half of the United States. Fishes obtained in California during these surveys were largely described by Charles F. Girard (1832-1895). Forty of the species authored by Girard belong to families covered in this booklet including the Pacific sardine, northern anchovy, white seaperch, Pacific tomcod, and a number of sculpins.

Between 1854 and 1875, one hundred five species were described and added to the California fish fauna. Louis Agassiz (1807-1873), William P. Gibbons (1812-1897), Friedrich H. von Kittlitz (1799-1874), Albert Günther (1830-1914), Theodore N. Gill (1837-1914), James G. Cooper (1830-1902), and Franz Steindachner (1834-1919), besides Ayres and Girard, described fishes that are still recognized today.

Two outstanding and distinguished ichthyologists dominated the California scene between 1880 and 1900. David S. Jordan (1851-1931) and Charles H. Gilbert (1859-1928) discovered and described thirty-five species during a three-year span, 1880-1883, while Gilbert himself described an additional twenty-three forms over a longer time. In the twenty-five-year span beginning in 1875, one hundred six species were discovered and described. Although Jordan and Gilbert were the major contributors, nine other prominent scholars were active during this period, including: Tarleton H. Bean (1846-1916), Frank Cramer (1861-1948), Carl H. Eigenmann (1863-1927) and his wife Rosa Smith (1859-1947), Barton W. Evermann (1853-1932), Samuel Garman (1843-1927), Arthur W. Greeley (1875-1904), William N.

Lockington (1842-1902), and Edwin C. Starks (1867-1932).

California's food and game fishes were brought into a reasonable degree of order and completeness with publication of the *Fishes of North and Middle America* by Jordan and Evermann at the turn of the century. Since 1900, twenty-three species have been discovered and added to the California fauna. The most recently described forms among the species included in our checklist of food and game fishes are two rockcods, *Sebastes reedi* described by S. J. Westrheim and H. Tsuyuki in 1967, *S. phillipsi*, recorded by John E. Fitch in 1964, and a clupeid, *Opisthonema medirastre*, by Frederick H. Berry and Izadore Barrett in 1963.

Three California institutions merit special consideration in this historical account. Stanford University was for years the great ichthyological center. Led by its president, David S. Jordan and his legendary array of co-authors, colleagues, and graduate students, it has until recently produced some of our country's most talented ichthyologists, but it ceased work in the field of fish taxonomy in 1970. Scripps Institution of Oceanography, which began in 1892 as a field investigation center and was named in 1925, has been a major influence in the California marine sciences, particularly with respect to fishes. Finally, the California State Fisheries Laboratory, established in 1917 with only one fishery biologist, Will F. Thompson (1888-1966), has grown into a major institution and is represented along the entire California coastline by separate laboratories, seagoing research vessels, and a highly competent staff.

CALIFORNIA'S FOSSIL RECORD OF FOOD AND GAME FISHES

Most, if not all, of the more than 525 kinds of marine teleost fishes known to California have been living off our shores for upwards of 12 million years, and a few

[13]

were here 25 million years ago. Identifiable otoliths and teeth of more than 150 different kinds of fishes (not counting sharks, skates, and rays) have been found in marine Pliocene and Pleistocene deposits throughout California, and there are mackerel skeletons, halibut otoliths, and sheephead teeth that are identical to their living counterparts in Miocene exposures.

The fish remains in many of these outcrops reflect oceanographic conditions at the time of deposition. The Palos Verdes Sand, which is the youngest of our marine Pleistocene (perhaps 120,000 years old), is highly fossil-iferous and contains shells, crab claws, sand dollars, sea urchin spines, barnacles, and many other invertebrate fragmentia as well as fish remains (otoliths, vertebrae, stings, and teeth). At least six kinds of otoliths (four croakers, a cusk-eel, and a searobin), found in exposures of Palos Verdes Sand at Playa del Rey, San Pedro, and Newport Beach, came from fishes that have never been observed north of Magdalena Bay during modern times. Finding remains of these semitropical species in south-ern California Pleistocene deposits is solid evidence that our ocean temperatures were considerably warmer 120,000 years ago than now.

On the other hand, an exposure of Lomita Marl (late Pliocene estimated to be older than 3 million years) at San Pedro yielded otoliths of nearly ninety species, and six of these were from locally extinct northern fishes. Two of these six have not been caught south of British Columbia during modern times, and the other four are not known to range much south of Eureka. Obviously the ocean at San Pedro was much colder when the Lomita Marl was being laid down than it is today. The Palos Verdes Sand has never yielded remains from "northern" fishes, nor has the Lomita Marl contained identifiable material from "southern" fish species.

Only one of the more than one hundred fifty kinds of marine teleost fishes that left their remains in Pliocene and Pleistocene deposits is now extinct. Ninety-one kinds are from families included in this book, with seventy of the ninety-one belonging to just six families: rockcods (17), sculpins (15), two flatfish families (14), perch (12), and croakers (12). Two of the twelve croakers live in waters south of California and a third is extinct, but the other nine are year around residents off our coast as are sixteen of the seventeen rockcods, fourteen of the fifteen sculpins, and all of the perch and flatfish. One rockcod and one sculpin inhabit waters north of California.

Pliocene and Pleistocene deposits that are rich in fish remains, primarily otoliths, are rapidly being destroyed by highway construction, housing developments, cut-and-fill projects, and other digging and earth-moving activities. Occasionally these projects uncover new exposures and these often yield valuable information on faunal associations during the most recent 12 million years of California's history.

In southern California, there are good marine Pliocene and Pleistocene exposures near the Mexican border, inland from Newport Beach, throughout much of the area between Long Beach and Playa del Rey, and in and around Ventura and Santa Barbara. In central and northern California, shelly outcrops near San Luis Obispo, Santa Cruz, and Arcata contain fish otoliths, teeth, and vertebrae.

Miocene deposits east of Bakersfield are also rich in otoliths, teeth, and vertebrae, but only teeth and vertebrae have been found in Miocene strata south and east of Santa Ana and on San Clemente Island. Skeletal imprints of tunas, flatfish, herrings, sea bass, jacks, and other families described here are abundant in Miocene diatomites and shales, particularly north of San Clemente, in the Palos Verdes area between San Pedro and Redondo Beach, in the hills between Anaheim and Pomona, in the Santa Monica Mountains, and around Lompoc. Miocene fossils are believed to be 12 to 25 million years old or older.

[15]

Oligocene deposits east of Bakersfield have yielded otoliths and teeth from nearly thirty kinds of marine teleosts that were living there some 25 to 36 million years ago. Otoliths of extinct bonefish, sea catfish, eels, and many other fishes are abundant in Eocene strata north and east of San Diego, and they could be nearly 60 million years old.

As previously mentioned, coastal Indian middens have yielded information regarding fish species utilized for food during the 9,000 years or so the Indians inhabited our shores. In addition these middens have also contained the remains of fishes that had been preyed upon by mammals, birds, and fish that were harvested by the Indians. Unfortunately, Indian middens are disappearing at an even more rapid rate than fossil deposits, and for the same reasons, namely modern man's digging and earth-moving activities — sometimes called "progress" for want of a better term.

Although the Los Angeles County Museum of Natural History contains the largest collection of the state's fossil fishes and fish remains, an excellent assortment also is housed at the California Academy of Sciences, San Francisco. Most of the state's colleges and universities have a few fossil shark teeth and diatomite "imprints" on hand for use in teaching paleontology, but neither these nor similar material that finds its way into high school biology classrooms are on public display.

ACKNOWLEDGMENTS

Many individuals shared their knowledge, ideas, or talents with us in our efforts to produce a factual, yet readable, account of the families of food and game fishes inhabiting our waters. We are extremely grateful for the assistance rendered by these friends and colleagues and hope we have not overlooked anyone in thanking them by name. In alphabetical order they are: John L. Baxter, E. A. Best, John Bleck, William L. Craig, Charles Crawford, Lillian Dempster, W. I. Follett,

Spano Giacalone, Daniel W. Gotshall, Roberta Greenwood, Carl L. Hubbs, Andrew Kier, Leo Pinkas, Jay C. Quast, Jack W. Schott, Norman Wilimovsky, Parke H. Young, and Louis Zermatten.

We feel, as do many, that it is the illustrations which "make" a book such as this, and we are fortunate to have had the assistance of the artistic and photographic talents of Arthur Bryarly, Harold Clemens, Charles Crawford, Daniel W. Gotshall, Evie Templeton, and Charles H. Turner.

Dorothy Fink, Betty Ponti, and Loretta M. Proctor typed various sections and drafts of the manuscript, and Arline Fitch graciously read proof on several occasions and offered helpful comments for improving several chapters.

FAMILY ACCOUNTS
Acipenseridae (Sturgeon Family)
Green Sturgeon
Acipenser medirostris Ayres, 1854

Distinguishing characters.—The ventral mouth which has four barbels in front of it, the narrow elongate snout, and the twenty-three to thirty bony shields in the lateral row (midside) will distinguish a green sturgeon from all other fishes in our waters.

Fig. 1.   *Acipenser medirostris*

Natural history notes.—*Acipenser medirostris* normally inhabits estuarine waters and river systems along the Pacific Coast from Alaska to San Francisco, but individuals occasionally enter the ocean where they have been caught as far south as Ensenada, Baja California. They are reported to reach a length of 7 feet and a weight of "approximately 350 pounds," but those caught in the ocean seldom exceed 3½ feet or 9 pounds.

[17]

One caught off Point Conception was 37 inches long and weighed 7¾ pounds, whereas a 36-inch fish hooked from the Belmont Pier (Long Beach) weighed 6½ pounds, and a 35-incher weighed 5¼.

No life-history studies have been made on the green sturgeon, but 138 individuals caught in gill nets in the Sacramento-San Joaquin Delta during 1963 and 1964 appeared to represent two different ages (possibly one and two years old). These sturgeon averaged between 10 and 18 inches long to the fork of the tail. Judged by these standards, a 3-foot sturgeon would be about seven years old.

Young fish, such as those caught in the Delta, feed almost exclusively upon crustaceans (amphipods, mysids, and other shrimplike creatures). Large individuals will eat shrimp, clams, and fish, but their piscivorous diet consists primarily of fishes that are small, disabled, or dead. We do not know of any specific predators on green sturgeon.

Acipenser remains (bony shields) have been found in Indian middens near some of the larger rivers along the Pacific Coast, and some of these probably have been from green sturgeon. Bony shields have also been found in fossil deposits, but these cannot be identified with *A. medirostris*.

Fishery information.—Between 1918 and 1964, it was illegal to "take or possess" sturgeon in California waters, but since then, a sport fishery has been authorized for fish exceeding 40 inches in length. This fishery is conducted almost exclusively in the Delta area and tidal channels connecting to San Francisco Bay, and perhaps 90 percent of the catch is composed of white sturgeon. Through 1968 the peak catch by partyboat fishermen was 2,200 sturgeon (in 1967). Possibly several hundred additional fish are taken each year by nonreporting anglers fishing from shore, and private skiffs and boats. The green sturgeon is considered inferior to the

[18]

white sturgeon as table fare, but both are very desirable.

There is no commercial fishery for sturgeon in California, and has not been since 1918. During the two years prior to the closure (1916 and 1917), 15,000 and 10,000 pounds, respectively, were marketed. Almost all of this poundage consisted of white sturgeon.

Other family members.—The white sturgeon is the only other member of the family known to California. It too strays into the ocean upon occasion and has been captured as far south as Ensenada. The two sturgeons are easy to distinguish by their head shape, color, position of the barbels in front of the mouth, and scale counts. Since counting the number of scales in the lateral row is a positive way to identify the two, this character is used in the following key.

1. If there are 23 to 30 bony shields in the lateral row (along the midside), it is a green sturgeon.
2. If there are 38 to 48 bony shields in this row, it is a white sturgeon.

Meaning of name.—*Acipenser* (ancient name for sturgeon) *medirostris* (moderate snout).

Albulidae (Bonefish Family)
Bonefish
Albula vulpes (Linnaeus, 1758)

Distinguishing characters.—The single dorsal fin at the middle of the back, the cone-shaped snout that projects beyond the mouth, the smooth scaleless head, and the silvery body covered with relatively small scales will distinguish the bonefish from all other fishes in our waters. Freshly caught individuals have faint horizontal streaks and dusky vertical bars along the sides, but these fade upon death. Fin bases on the under side of the fish are yellowish.

Natural history notes.—*Albula vulpes* is reported in the literature as a circumtropical species the adults of which have been recorded from all tropical seas. In the eastern Pacific, *A. vulpes* has been taken at many localities between San Francisco Bay (once) and Panama. A world record (for hook and line) bonefish 3 feet 3-5/8 inches long, weighing 19 pounds was caught off Zululand, South Africa, in 1962, and individuals weighing up to 22 pounds have been netted elsewhere in the world. A large bonefish in the eastern Pacific may never exceed 2 pounds in weight or 18 inches, and for this and other reasons (namely differences in otolith configuration and proportions) we believe that several rather than a single species of *Albula* are living throughout the world. A 1½-pound female taken at Magdalena Bay was 17½ inches long, and judged by growth rings on its otoliths, seven years old.

Fig. 2.   *Albula vulpes*

It is believed that spawning occurs offshore in deep water, and that the eggs are pelagic, but there is no proof of this. Information is lacking on size and age at maturity, fecundity, and all other facets of reproduction.

Bonefish larvae have a forked tail fin but otherwise resemble the transparent larvae of eels, and are known as leptocephali. For a short period, a bonefish leptocephalus grows (lengthens), but when it reaches about 2½ inches it starts to develop fins and it commences shrinking. The shrinking process continues for ten or twelve days until the leptocephalus is only about half its original length. It then metamorphoses (takes on a bonefish appearance for the first time), and once again starts growing.

Clams and small snails made up over half of the diet of bonefish studied in Puerto Rico. Crabs comprised one-third of the food eaten, and shrimp and small fish completed the diet. Those we have examined from our coast had been eating clams, crustaceans (mostly small crabs), and fish. We have no information on bonefish predators.

Fossil albulids, not *Albula vulpes*, have been noted from many parts of the world based upon otoliths, dentition, and skeletal fragments. Some of these fossils date back to the Cretaceous (about 125 million years), but the greatest number is found in the Eocene. There are otoliths of fossil albulids in Eocene deposits near San Diego, California.

Fishery information.—Except for a dozen or so individuals taken in bait nets by commercial fishermen and several seined by fishery biologists during scientific investigations, all the bonefish from California have been caught by sport fishermen. They occur in our waters only sporadically, and seldom are more than two or three taken during any given year. Some have been caught from piers, but most have been caught in or near shallow bays and harbors of southern California. The bonefish of the eastern Pacific doesn't grow large enough to put up a very good fight, even on light tackle. Its qualifications as table fare are questionable; its flesh is filled with small bones.

Other family members.—No other member of the family is known within our area.

Meaning of name.—*Albula* (white) *vulpes* (fox).

Anarhichadidae (Wolffish Family)
Wolf-eel
Anarrhichthys ocellatus Ayres, 1855

Distinguishing characters.—The lack of pelvic fins, extremely elongate tapering body, slender and almost indistinguishable caudal fin, large conical canine and molarlike jaw teeth, and distinctive color (large ocellated black spots) will separate the wolf-eel from all other fishes in our waters. The wolf-eel is closely related to the blennies; it is not an eel.

[21]

Natural history notes.—Anarrhichthys ocellatus ranges from Kodiak Island, Alaska, to Imperial Beach, California, but it is not common south of Point Conception. In the cold northern waters, it inhabits relatively shallow rocky areas, but to the south it is found at much greater depths. One taken near La Jolla was hooked in 400 feet of water, and others have been caught off southern California in depths greater than 250 feet.

Fig. 3. *Anarrchichthys ocellatus*

Various accounts report that they attain a length of 8 feet, but this is obviously an estimate made by some writer several decades ago, and subsequent authors have "followed the leader." The largest authentic wolf-eel appears to be a 6-foot 8-inch fish speared off Rosario Beach, Washington, in 1962; this fish weighed 40 pounds 10 ounces. Three smaller wolf-eels (38, 51, and

[22]

61 inches long) weighed 2½, 7, and 12-1/3 pounds, respectively. Based upon these lengths and weights, an 8-footer would have to weigh over 100 pounds, and possibly as much as 150 pounds. An immature male that was 35 inches long appeared to be four years old, judged by some rather vague growth rings on its otoliths.

Spawning takes place during winter months, and the whitish eggs are deposited in a mass on a protected surface of a rocky cave or crevice. They adhere to the rocky substrate, and both parents remain on guard until hatching takes place. One nest contained an estimated 7,000 eggs, but information is lacking on the size of the female. We have no information on age or length at first maturity, number of eggs for a female of any given size, time required to hatch, or other facets of reproduction.

The wolf-eel stomachs we have examined have contained a preponderance of crab remains. Other items we have observed were sea urchin fragments, small snails including abalones, and an occasional piece of fish. There is a report of a 16½-inch wolf-eel having been found in a salmon stomach, but we do not know of any other predation.

Fishery information.—Sport fishermen take an estimated 200 wolf-eels a year, according to a survey conducted from 1957 to 1961. About two-thirds of these are speared by skindivers, while most of the remainder are caught by skiff fishermen. Those taken on hook and line usually have been attracted by an anchovy, some other small bait fish, or a piece of abalone.

There is no commercial fishery for wolf-eels.

Other family members.—No other member of the family is known in the eastern Pacific Ocean.

Meaning of name.—*Anarrhichthys* (*Anarhichas* fish, for its resemblance to an Atlantic relative) *ocellatus* (with eyelike spots).

Anoplopomatidae (Sablefish Family)
Sablefish
Anoplopoma fimbria (Pallas, 1814)

Distinguishing characters.—The general color of the sablefish (blackish or grayish on top shading to lighter below), and the shapes and placement of the various fins will serve to distinguish it from all other fishes in our waters. The jaw teeth are hardly noticeable, and because of the rounded snout, the mouth somewhat resembles that of a frog. The flesh is fine-grained and very oily.

Fig. 4.   *Anoplopoma fimbria*

Natural history notes.—*Anoplopoma fimbria* ranges from the Bering Sea to Cedros Island, Baja California, at least. The blue-colored larvae and early juveniles live at the surface, often many miles off shore, but young fish (to a foot long or more) are found on the bottom in relatively shallow water. Adults also live on the bottom and prefer areas of blue clay or fairly firm mud. At the more southerly latitudes they live in deeper water than in the northern portion of their range. Captures have been made at depths exceeding 5,000 feet off southern California and northern Baja California, and they appear to be abundant at these depths.

A record sablefish is said to have weighed 56 pounds, but during the past several decades few individuals have been seen that weighed as much as 30. A 30-pounder will be about 40 to 42 inches long, and twenty years old or older. About 50 percent of the male sablefish are mature when they average about 23½ inches in length and five years in age, but females take an additional year to mature and are 4 inches longer. These fish will average 4 and 6½ pounds, respectively, and the 27½-inch female will produce about 100,000 eggs, compared to about 1 million for a 40-inch female. Females reach greater sizes and ages than males.

Spawning occurs in late winter and early spring, and in the northern part of their range the adult population moves into deeper water to spawn. Their behavior in more southerly climes is not known. The eggs are said to be pelagic, and they drift at or near the surface until they hatch. Aside from their spawning migrations, sablefish seldom move any great distance during their life. Tagging studies showed that 30 miles is about the maximum range of an individual in four or five years.

Adult sablefish feed primarily upon fish, squid, octopi, shrimplike crustaceans, worms, and an assortment of other small organisms that live in or on a mud or clay substrate. Small sablefish are fed upon by numerous predators including albacore, rockcod, salmon, lingcod, seals, and sea lions during their pelagic and shallow-water stages, but adults appear to be relatively free of predation except by large deep-living sharks, sperm whales, and hagfish.

Fishery information.—During some years, young sablefish abound in inshore areas, and thousands will be caught by pier and skiff fishermen in central and northern California waters. Because of the great depths they inhabit, there is no sport fishery for adults (few fishermen have the time or inclination to drop a baited hook to the bottom in a thousand feet of water and then reel in their catch, or an empty hook, when a sablefish bites).

The bulk of the commercial catch is made with otter trawls and multi-hook setlines. Although the catch has fluctuated since 1942 (it peaked at 6 million pounds in 1945 and hit a low of 900,000 in 1947), annual landings have averaged about 2 million pounds, and have been rather stable since 1950. A major portion of the catch is smoked, and because of the oily flesh it makes a superior product when so processed.

Other family members.—The skilfish, which is known to reach a length of 6 feet and a weight of 200 pounds, is the only other member of the family known from our waters. It is rarely seen, but easily distinguished from the sablefish by color (large white mottled areas on the sides, especially in younger individuals), larger size, and several external characters.

1. If there is a broad separation between the 2 dorsal fins, it is a sablefish.
2. If the 2 dorsal fins are closely approximated, it is a skilfish.

Meaning of name.—*Anoplopoma* (unarmed operculum) *fimbria* (fringed).

Ariidae (Sea Catfish Family)
Chihuil
Bagre panamensis (Gill, 1863)

Distinguishing characters.—This typical catfish is characterized by having only one pair of barbels on the lower jaw, and by the lack of an elongate filament on the dorsal fin spine. The first dorsal fin spine and similar-sized spines on each pectoral fin are barbed along one margin and are hollow. Venom, produced in a gland at the base of each spine, travels through the hollow center to the tip, and can cause a sickeningly painful wound to anyone unfortunate enough to be stuck by one. Pain is immediate and excruciating, and it is not restricted to the area of the wound. In extreme cases there is severe swelling of the affected member, dizziness, and nausea. Blood pressure and pulse rate often are altered for a short period.

Natural history notes.—*Bagre panamensis* has been recorded from near Newport Beach, California (on November 3, 1965), to south of Panama. It apparently has never been taken between Newport Beach and Magdalena Bay, but it is extremely abundant in Magdalena Bay, throughout much of the central and southern Gulf

[26]

of California, and south to Panama. It inhabits shallow bays and quiet offshore waters where the bottom is muddy or some combination of sand and mud. The chihuil has been reported as reaching a length of 18 inches, but none of the twenty-six or more kinds of sea catfishes inhabiting the tropical eastern Pacific has been studied in any detail, so this is not a firm size estimate. The Newport Beach catfish was just over 12 inches long and weighed about 9 1/2 ounces. Otoliths of *B. panamensis* have excellent growth zones on them, and these indicate that the species first matures when three or four years old, and some individuals will reach ages of fifteen to eighteen years.

The males of many sea catfishes brood the marblesized eggs in their mouths. Obviously, the fish cannot eat until the eggs hatch and the youngsters depart. The reproductive behavior of *B. panamensis* is unknown, but it very likely has similar habits.

Sea catfish feed primarily upon small fishes, whether they are dead or alive. They also will eat shrimp, small octopi, squid, and other soft bodied invertebrates that live in their environment. Juvenile and adult sea catfish are preyed upon by several kinds of sharks, especially hammerheads. Apparently these sharks are immune to the venomous catfish spines, which often become embedded in their jaws.

Otoliths of sea catfishes have been found in Eocene and Cretaceous deposits in California and some of the states bordering the Gulf of Mexico, but these are not from *Bagre*.

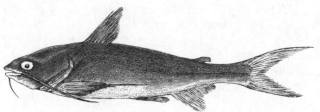

Fig. 5   *Bagre panamensis*

[27]

Fishery information.—Where sea catfishes are abundant, it is next to impossible to catch anything else when fishing with a baited hook. *B. panamensis* is not considered to be of very high quality as table fare, and seldom is utilized, even though taken in great quantities by Mexican shrimp trawlers and beach-seine fishermen.

Other family members.—No other sea catfish is known from north of the Magdalena Bay area.

Meaning of name.—*Bagre* (Portuguese name for some catfish) *panamensis* (Panama, the area of first capture).

Atherinidae (Silverside Family)
California Grunion
Leuresthes tenuis (Ayres, 1860)

Distinguishing characters.—The shape, size, color, and fin placement of the California grunion will distinguish it as a member of the silverside family. Its spawning behavior is unique among California's marine fishes in that it strands itself to deposit its sex products in moist sand. The absence of jaw teeth further differentiates it among the silversides that inhabit our waters.

Fig. 6   *Leuresthes tenuis*

Natural history notes.—*Leuresthes tenuis* has been recorded from Monterey Bay to San Juanico Bay, Baja California, but during the most recent quarter century it has seldom been seen north of Morro Bay, and is not observed at Morro Bay every year. When not spawning in the intertidal zone on wave-swept sandy beaches, grunion are scattered throughout shallow waters outside the breaking surf. They seldom are seen or captured more than a mile offshore or where water depths exceed about 60 feet.

[28]

Female grunion grow to larger sizes than males and reach a maximum length of about 7 inches. A 7-inch long fish will weigh less than 2 ounces, even when full of eggs. The spawning season extends from late February or early March to late August or early September, and fish that are 5 inches long in their first year of life are capable of spawning. Maximum age appears to be four years, but few fish live longer than three.

Females spawn about four to eight times during a season, at approximately two-week intervals, and release upwards of 3,000 eggs per spawning. The eggs remain buried in the moist sand for about ten days, and hatch when the next series of high tides digs them out and sloshes them around in the surf. If they fail to be dug out on the first series of high tides after spawning they will still be viable during the next series of high tides, perhaps twenty-two to twenty-five days after spawning.

While buried in the sand grunion eggs are fed upon by sand worms, probed for by shore birds, and even dug out by ground squirrels. After they hatch, and throughout their lives, grunion are fed upon by halibut, bass, white croakers, and other large predators. Grunion food habits are not as well known as other facets of their life history, but the stomachs examined have contained mostly microscopic and slightly larger planktonic organisms.

Fossil grunion otoliths have been found, although never abundantly, in several southern California Pliocene and Pleistocene deposits, so they have been off our shores for 10 to 12 million years, at least. Coastal Indians apparently harvested grunion during spawning runs, if the otoliths found at various Indian campsites can be used as a criterion for judging their fishing habits.

Fishery information.—Licensed sport fishermen are allowed to catch grunion only by hand and only during the open season. Spawning runs occur only on sandy beaches (often inside harbors and breakwaters), only at night, only for three or four nights following a full or new moon, and only for a several-hour period immediately after high tide. Predicted dates and times of runs are always listed in newspapers published in southern California coastal communities and in some metropolitan papers. A flashlight is useful for the grunion hunter, especially during the new moon runs. A successful hunter is going to get wet, so he should dress accordingly. Flashing lights seldom bother spawning grunion, but a beachful of fish can be emptied in a matter of seconds by stomping one's feet in the moist sand. For this reason, grunion hunters should never run or hurry when searching for or trying to catch some grunion.

The commercial catch of grunion probably seldom exceeds a few hundred pounds during any given year. The few that are taken incidental to other fisheries are marketed fresh and appear in catch statistics as part of the state's "smelt" landings.

Other family members.—Two other kinds of silversides are known from off California, and the only infallible way to identify all three species requires a magnifying glass (or microscope).

1. If there are no teeth in the jaws it is a grunion.
2. If the jaw teeth are set in a single row and are forked at their tips it is a topsmelt.
3. If the jaw teeth are set in bands and are small and simple (unforked) it is a jacksmelt.

Meaning of name.—*Leuresthes* (pertains to the toothless jaws) *tenuis* (slender).

Balistidae (Triggerfish Family)
Finescale Triggerfish
Balistes polylepis Steindachner, 1876

Distinguishing characters.—The tough leathery skin covered with small, solidly affixed scales; the tiny gill opening above and in front of the pectoral fin; the small terminal mouth filled with close-set (rabbitlike) teeth

[30]

which have moderately strong cusps; and the three-spined first dorsal fin which can be locked into an erect position are more than sufficient characters to identify the finescale triggerfish.

Fig. 7.   *Balistes polylepis*

Natural history notes.—*Balistes polylepis* is endemic in the eastern Pacific, and has been captured between Crescent City, California (once), and Afuera, Peru. It is abundant throughout the Gulf of California and south through tropical waters, but its occurrence off California is sporadic and unpredictable. The early juveniles are pelagic (living at the surface and well offshore), but the adults prefer inshore areas where the bottom is sandy, or where sandy and rocky habitat intermix. They do not shy away from rocky bottom areas, however.

We have seen finescale triggerfish that were about 30 inches long and weighed in excess of 10 pounds, and they have been reported in the literature as weighing up to 16 pounds, but vital statistics apparently have never been taken on the very largest individuals. An 18-inch fish taken off Paradise Cove weighed 3½ pounds,

[31]

but we have no information on the age or sex of this individual.

As common as this fish is in the Gulf of California, its life history has not been studied. There is no information on its age, growth, maturity, fecundity, or behavior. The few observations on its food habits are casual at best. Triggerfishes have strong jaws and teeth, and are noted for their ability to bite off bits of coral, and other sedentary animals that are encased in various kinds of "armor." We have found remains (shell fragments) of barnacles, clams, and snails in the stomachs of finescale triggerfish, as well as fish flesh. This is the only fish we know of that appears to relish shark meat (aside from a larger shark). In the southern Gulf of California, swarms of finescale triggerfish will be chewing on any shark carcass that is thrown overboard before it has time to reach bottom. They continue "attacking" the carcass until nothing remains.

Pelagic juveniles are fed upon by tropical tunas, wahoos, and other large predators that inhabit offshore waters, but we do not know of anything that preys on large adults.

Triggerfish teeth, possibly those of *B. polylepis*, have been found in Miocene fossil deposits near Bakersfield and Santa Ana that may be 15 to 25 million years old.

Fishery information.—There is no sport fishery for finescale triggerfish in California. The twenty or more individuals that have been reported from our waters have been hooked, speared, and netted. Only in the case of those that were speared did the fisherman make an effort to "catch" the species. Finescale triggerfish are excellent table fare, but they must be skinned, and because of their tough hide many fishermen become discouraged and discard them. In many parts of the world, triggerfish flesh is extremely toxic and can cause an agonizing death if eaten, even in small amounts. The finescale triggerfish has never been found toxic, however.

[32]

Other family members.—One other triggerfish has been reported from California, but only once. It is easy to identify these two triggerfish by color and body shape, but the characters noted below are easily observed (and explained).

1. If there are 4 to 6 conspicuous longitudinal grooves on the cheek below the eye, it is a redtail triggerfish.
2. If there are no grooves on the cheek, it is a finescale triggerfish.

Meaning of name.—*Balistes* (alluding to the trigger mechanism that comprises the spinous dorsal) *polylepis* (many scale).

Batrachoididae (Toadfish Family)
Plainfin Midshipman
Porichthys notatus Girard, 1854

Distinguishing characters.—The scaleless body which is purplish-bronze above and yellowish-white below, the large broad head with prominent widely separated protrusible eyes, and the numerous rows of luminous organs on the underside are sufficient to distinguish a midshipman from all other fishes in our waters. The plainfin midshipman is characterized by its unspotted dorsal, anal, and pectoral fins, and by the second row of photophores under the head which form a forward-directed V.

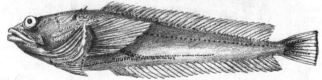

Fig. 8.　*Porichthys notatus*

Natural history notes.—*Porichthys notatus* is generally reported as ranging from Sitka, Alaska, to Magdalena Bay (by some authors) or the Gulf of California (by others). Those captured between about San Quintin Bay, Baja California, and Cape San Lucas, differ slightly from typical *P. notatus* of more northerly climes, but so

[33]

far they have not been considered as being a distinct species. They are abundant in the intertidal in cold northerly waters and to depths of 1,000 feet or more. They inhabit muddy bottom areas almost exclusively, and are noted for their habit of burrowing into the mud during daytime hours. At night they emerge from the bottom and move about in search of food.

The plainfin midshipman is reported to reach a length of 15 inches. The largest of some ten thousand or more individuals that we have seen was a 13½-inch male that weighed just over 14½ ounces. Male midshipmen, unlike most other fish species, grow larger and live longer than females. Their otoliths are very dense and growth zones usually are impossible to find, but sufficient good otoliths can be found to indicate that females reach an age of at least three years, while some males reach four.

Spawning takes place during late spring and early summer, and the large yellowish eggs (some are nearly ¼ inch across) are deposited on the protected side or undersurface of rocks, shells, or other solid substrate, where they adhere tightly in a single layer. A large female will spawn around 200 eggs, and the male midshipman tends these eggs until they hatch, guarding them against predation and flushing sediment from them periodically. Most fish of both sexes are believed to die after spawning. Midshipmen use their luminescent organs during courtship display, but at other times it is very difficult to make one light up. A sharp opercular spine on each side of the head can cause much pain, but it does not appear to be venomous or toxic.

Midshipmen will eat just about any type of organism they can get their mouth around, whether it is alive or dead. Small shrimplike crustaceans, and fish (mostly anchovies) appear to make up the bulk of their diet, but more than a dozen other kinds of food items have been noted. Plainfin midshipmen have been found in the stomachs of rockcod, giant sea bass, lingcod, sea lions, and numerous other large predators.

Otoliths of *Porichthys notatus* often are abundant in Pliocene and Pleistocene deposits throughout California. Some of these fossil otoliths are estimated to be 12 million years old or older. Their otoliths are commonly encountered in coastal Indian middens, but we do not know whether they were caught and eaten by the Indians or were in the stomachs of sea lions the Indians killed for food.

Fishery information.—There is no sport fishery for midshipmen, although a few are caught incidental to other hook-and-line fisheries each year, and skindivers spear occasional large individuals.

The commercial catch is made almost exclusively with trawling gear incidental to other fisheries. Most of these are thrown back along with other "trash" species, but many are retained and canned for pet food or frozen for use on fur farms. There is no reason why they could not be marketed for human consumption; they are rather free of parasites and their flesh is white, fine-grained, and mild.

Other family members.—One other midshipman inhabits our waters. It attains a larger size (to 18 3/4 inches and 3 pounds), is found in shallower depths, and prefers warmer more southerly climes.

The two species are easily identified.
1. If the dorsal, anal, and pectoral fins are heavily spotted, it is a specklefin midshipman.
2. If these fins are without spots it is a plainfin midshipman.

Meaning of name.—*Porichthys* (pore fish, with reference to the well-developed mucous system) *notatus* (spotted, pertaining to the photophores).

Belonidae (Needlefish Family)
California Needlefish
Strongylura exilis (Girard, 1854)
Distinguishing characters.—The greatly produced jaws which are filled with needle-sharp teeth are suffi-

cient to distinguish a California needlefish from all other species in our waters. If confirmation is needed, other distinctive characters include dorsal and anal fins set far back; the very long, almost round body; and the body color—greenish above and grading to silvery-white below.

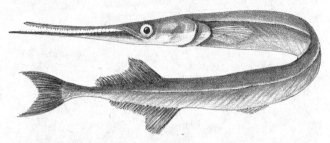

Fig. 9.   *Strongylura exilis*

Natural history notes.—Strongylura exilis has been recorded from San Francisco to Peru, but there are only a couple of occurrences north of Santa Barbara. When seen in our waters, they usually are in small "schools" (perhaps five to fifteen individuals) at or near the surface in bays, sloughs, or harbors. They are not a fish of the open ocean. They are said to reach a length of 3 feet, and probably do, but the largest we have measured was a 30-inch female that weighed just under 1 pound. The otoliths of this fish, caught at Redondo Beach in September 1960, had five winter rings, indicating an age of five years. A 19½-inch male taken in March 1960 at Redondo Beach weighed less than 4 ounces and was only one year old.

Spawning apparently takes place in the spring months, but we have no information on where they spawn, age at first maturity, number of eggs spawned each season by a single fish, or other details of reproduction. One interesting feature of the California needlefish is that only one gonad (the right one) is ever developed or functional. There is no ready explanation for this phenomenon.

Many needlefish are heavily parasitized, both internally and externally. One caught off Carpinteria in September 1965 had six plumelike copepods (*Pennella*) sticking out of its sides. The other ends of these hairy-looking parasites were firmly and permanently attached to large blood vessels inside the fish.

Small fishes, usually anchovies, are the only food items we have found in their stomachs. We know of no predators on the California needlefish, and there is no fossil record of the species.

Fishery information.—The California needlefish is neither a commercial nor a game fish, although a 30-inch fish on very light tackle will put on quite an aerial display. They have been caught accidentally (or incidentally) during every month of the year, but not necessarily during any given year. Several decades ago dozens of them were speared each year from the bridge that crossed Mission Bay (San Diego County) between Ocean Beach and Mission Beach. Since the bridge was removed, most catches have been made with hook and line, or in roundhaul (bait) nets.

Other family members.—No other member of the family is known from closer than the Gulf of California.

Meaning of name.—*Strongylura* (round tail) *exilis* (slender).

Bothidae (Lefteye Flounder Family)
California Halibut
Paralichthys californicus (Ayres, 1859)

Distinguishing characters.—The eyes of this flatfish are both on the right side in about 48 percent of the individuals, even though it belongs to the lefteye flounder family. Whichever side the eyes occur on, the California halibut is easily distinguished from all other flatfish by a combination of three characters: a large mouth, numerous sharp teeth, and a high arch in the lateral line above the pectoral fin. On rare occasions California halibut are seen that are brown on both sides (ambicolored), or even more rarely, white on both sides (albinistic).

Natural history notes.—Paralichthys californicus has been captured from Alsea, Oregon, to Magdalena Bay, Baja California, but it shows up only seasonally north of Morro Bay, and sporadically north of Tomales Bay. It is most abundant on sandy bottoms in water shallower than 120 feet but is known to inhabit depths to 300 feet or deeper. The California halibut is the larg-

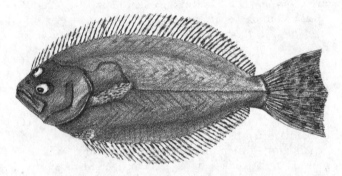

Fig. 10. *Paralichthys californicus*

est of fifteen or more kinds of *Paralichthys* known from both coasts of North and South America. There is an unverified report of a 72-pounder that was said to be 5½ feet long, but the accepted record is a slightly shorter 61½-pound fish. Any individual weighing over 50 pounds is worthy of note, however. Spawning takes place in relatively shallow water from about February through July, and in two or three years male halibut are mature but the females do not spawn until they are four or five. A five-year-old fish will average about 15 inches long. Females grow faster and attain larger sizes than males, and almost all fish weighing over 30 pounds are females. The oldest (age thirty) of several hundred individuals examined in a recent study was not the largest;

[38]

this fish, a female, was 45 inches long and weighed 40 pounds, whereas a 50-inch long female weighing 51 pounds was only seventeen years old.

A tagging study showed that *P. californicus* was not a traveling fish, especially when young. Of more than 12,000 fish tagged during a several-year period, only about 10 percent of those recovered had traveled more than a mile from the point of release. The greatest distance one moved was 140 miles during a 527-day period after tagging.

Halibut feed almost exclusively on anchovies, queenfish, and similar small fish, and at times they can be seen jumping clear of the water as they make passes at anchovy schools near the surface. Halibut and halibut remains have been found in the stomachs of angel sharks, electric rays, sea lions, Pacific bottlenose dolphins, and several other predators.

Halibut remains are abundant in Indian middens along the southern California coast, some of which were occupied more than 5,000 years ago. *Paralichthys* otoliths have been found in Miocene deposits near Bakersfield that may have been laid down 25 million years ago, but these may not be from *P. californicus*.

Fishery information.—Sport fishermen do most of their fishing in water shallower than 60 feet, by drifting or slow trolling with both live and dead bait or a shiny metal lure. Large lively anchovies or medium-sized queenfish (5- to 7-inchers) are the choicest baits, but shiner perch, small señoritas, herring, sardines, and a few other species will also attract halibut. North of Point Conception, best fishing is found inside Morro Bay, in Moss Landing Harbor, outside the breakers at Aptos and Seaside, just outside the entrance to San Francisco Bay, and at Tomales Bay, but runs of good fish usually are restricted to a few months each year. South of Point Conception, fishing is best outside the surf zone where the bottom is sandy. Areas that have yielded fair to good catches consistently, but not necessarily the

[39]

only such localities, are: the Santa Rosa Island area, Ventura flats, Zuma Beach, Long Beach Harbor, Huntington Beach flats, Oceanside, and Coronado Strand. Partyboat fishermen have caught as few as 10,000 halibut per year since the Second World War and as many as 150,000. Skiff fishermen, shore fishermen, and skindivers probably catch an equal number.

The greatest recorded commercial catch, just over 4½ million pounds, was made in 1919. Since then landings have fluctuated widely, dropping as low as one-fifteenth the peak catch during some years. Most of the commercial halibut catch is made with trammel nets and otter trawls and is sold to the consumer in the form of fillets. Marauding sea lions often wreak havoc with halibut they find in trammel nets. Fish they do not eat entirely are left in such ragged and unsightly condition that they cannot be marketed. Often an entire night's catch will be destroyed by one or two such sea lions. Halibut flesh is white, fine-grained, and very mild and makes excellent table fare regardless of how it is served.

Since there are several laws relating to the taking of halibut by both commercial and sport fishermen, one should always check current regulations before going fishing.

Other family members.—Five other lefteye flounders occur off California, and even though many California halibut and fantail soles have their eyes on the right side, the characters listed below are sufficient for identifying these as well as the more typical lefteyed individuals.

1. If there is a high arch in the lateral line above the pectoral fin on the eyed side, and
    a. if the pectoral fin is longer than the head, it is a fantail sole.
    b. if the pectoral fin is much shorter than the head, the maxillary (upper jaw) reaches past the rear margin of the eyes, and
        i. the mouth is full of strong, sharp teeth, it is a California halibut.

[40]

ii. the jaw teeth are weak and tiny, it is a bigmouth sole.
2. If the lateral line is straight for its entire length, and
   a. if the pectoral fin is longer than the head, it is a longfin sanddab.
   b. if the pectoral fin is much shorter than the head, and
      i. the eye is shorter than the snout, and there are nine or fewer rakers on the lower limb of the first gill arch, it is a speckled sanddab.
      ii. the eye is longer than the snout, and there are ten or more rakers on the lower limb of the first gill arch, it is a Pacific sanddab.

Meaning of name.—Paralichthys (parallel fish) *californicus* (Californian). A Californian fish that lies parallel.

Pacific Sanddab
Citharichthys sordidus (Girard, 1854)

Distinguishing characters.—The Pacific sanddab is easily identified among small-mouthed, lefteyed flatfishes by a combination of characters: a normal appearing tail (the tonguefish lacks a distinct tail); unbanded dorsal, anal, and caudal fins (starry flounders have black-banded fins and although they belong to the righteye flatfish family, many are lefteyed); a pectoral fin on the eyed side that is shorter than the head; eyes that are longer than the snout; and ten or more rakers on the lower limb of the first gill arch.

Natural history notes.—Citharichthys sordidus is reported to range from northwestern Alaska to the tip of Baja California, but they are not abundant south of about San Quintin Bay. Larvae are commonly taken in plankton and other fine-mesh nets towed near the ocean's surface many miles offshore, but juveniles and adults live on the bottom where the substrate is fairly firm sand or sandy mud. They are most abundant at depths of 120 to 300 feet but have been captured shallower than 60 feet and deeper than 600.

The Pacific sanddab is reported to reach 16 inches

[41]

and weigh 2 pounds, but individuals exceeding 12 inches and 10 ounces are rarely seen. Females grow faster than males and generally mature when about three years old and 7½ inches long. An 11½-inch long female was nine years old, judged by growth rings on its otoliths. Spawning takes place from July through September during most years, and each female probably spawns more than once per season. The eggs are so small (0.57 to 0.77 millimeters in diameter) it would take 35 to 45 of them in a line to make an inch. The sex of a sanddab can be told by holding the fish up to a strong light. If it is a female, the long tapering ovaries can be seen extending backward from the body cavity just above the ventral edge of the fish. Pacific sanddabs eat a wide variety of small fish, cephalopods, pyrosomes, crustaceans, marine worms, and other food items that fit their mouths. As planktonic larvae they are eaten by albacore, salmon, mackerel, and numerous other predators that feed in the upper water layers, while as juveniles and adults they supply nourishment to most large voracious fish that live on or near the bottom. Diving birds such as cormorants, seals and sea lions, and a few porpoises also take their toll of sanddabs.

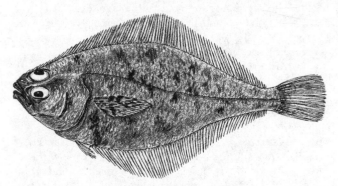

Fig. 11.   *Citharichthys sordidus*

C. sordidus otoliths have been identified from many marine Pleistocene and Pliocene deposits throughout the state, and some of these probably are 10 to 12 million years old. Miocene deposits near Bakersfield (possibly 25 million years old) contain *Citharichthys* otoliths, but not those of C. *sordidus*.

Fishery information.—Hook-and-line fishermen who want to catch sanddabs usually need only to seek the proper depth and bottom and use a tough bait on small hooks in order to get a good mess. Because of the time required to reel in from a depth of about 300 feet, most fishermen use a line that terminates in more than one hook. A favorite rig involves use of a large metal (wire) hoop around which a dozen or more small hooks have been tied. This is baited with small pieces of squid, clam, fish, or shrimp and lowered to just off the bottom. If sanddabs are present, only a few minutes are required to fill all the hooks. Consistently good areas are found near most skiff launching sites and harbors along the entire coast, but favorite areas are offshore from La Jolla, Newport Beach, Avalon, San Pedro, Malibu, Goleta, Pismo Beach, Pacific Grove, Moss Landing, and Capitola, to name a few.

The commercial catch, made almost entirely with otter trawls from the ports of Monterey, San Francisco, and Eureka, has held steady at about 500,000 to 700,000 pounds per year since 1930. The peak catch was made in 1916, when 2.2 million pounds were landed. Sanddabs are usually prepared for the table by removing heads, viscera, fins, and scales, and then deep frying them either with or without a covering of batter.

Other family members.—For information on the other lefteye flounders, including sanddabs, please refer to p. 40 and to our checklist (p. 159).

Meaning of name.—*Citharichthys* (a fish which lies on its side) *sordidus* (sordid). A dull-colored fish which lies on its side.

Bramidae (Pomfret Family)
Pacific Pomfret
Brama japonica Hilgendorf, 1878

Distinguishing characters.—The oblong, compressed body, deeply forked tail, long pectoral fins, and relatively small scales will distinguish the Pacific pomfret from all other fishes in our waters. The thin caudal peduncle, and long dorsal and anal fins are also useful recognition characters.

Fig. 12.   *Brama japonica*

Natural history notes.—*Brama japonica* is found throughout the temperate north Pacific Ocean, ranging on our coast from the Bering Sea to just north of Cedros Island, Baja California. It is a surface-dwelling form and is most abundant in northerly offshore waters. Although it has been reported as reaching a length of 4 feet, this is obviously an "educated guess," and does not represent an actual measurement. In our search for an authentic record size, we questioned fishery biologists who had captured thousands of pomfrets in the Gulf of Alaska and off the Aleutians, but these individuals had never measured nor weighed a really large specimen. The largest from off our coast was a 20-inch female that weighed nearly 3¼ pounds. An examination of the otoliths of this fish indicated it was six years old.

[44]

We were unable to find any information on size at first spawning, spawning season, or number and size of eggs per female. It is believed that spawning takes place in the south of the Pacific pomfret's range during the early spring, and that the spawned-out adults then migrate into the more northerly parts of their range. Larvae and juveniles are common in the offshore area south of about Point Conception to about the latitude of Magdalena Bay, Baja California, at least, but they are rare to the north.

Stomachs of several individuals from off California have contained primarily crustaceans (amphipods and euphausiids), but small fishes and squids are also common and probably contribute more bulk to the pomfret's diet than do the tiny crustaceans. Juvenile pomfrets, to 4 or 5 inches, have been found in lancetfish stomachs on many occasions, and an albacore will occasionally regurgitate one it has eaten. Sharks and marine mammals will eat larger pomfrets.

Fishery information.—Albacore fishermen and salmon trollers catch most of the pomfrets that are taken off California each year. Fair numbers are caught by crew members on our weather ships while they are on station several hundred miles offshore. In the Gulf of Alaska and off the Aleutians thousands of pomfrets have been caught in salmon gill nets and with purse seines. Some gill nets yielded as many as 232 pomfrets per set, while as many as 110 were taken in a single purse-seine haul. It has been suggested that a profitable long-line fishery very likely could be developed for the Pacific pomfret somewhere off central California during the late spring. Long-line gear is used successfully off the coast of Spain for the Atlantic pomfret.

Other family members.—Two other members of the family have been recorded from off California, but both are very rare. They are easily recognized as bramids by their body shape (laterally compressed and oval to ob-

[45]

long), and can be distinguished from each other with a casual glance.

1. If the dorsal and anal fins are very long and fanlike it is a fanfish.
2. If these fins are not fanlike and:
   a. if there are about 40 rows of scales along the mid-side it is a bigscale pomfret.
   b. if there are about 75 rows of scales it is a Pacific pomfret.

Meaning of Name.—Brama (a bream) *japonica* (Japan, the area from which the species was first described).

Branchiostegidae (Tilefish Family)
Ocean Whitefish
Caulolatilus princeps (Jenyns, 1842)

Distinguishing characters.—The ocean whitefish is easily identified by a combination of several characters: a very long dorsal fin that is nearly the same height throughout its length, a long anal fin with all rays of about equal length, and a brownish-green or olive-colored body covered with quite small scales. The elongate, tapered body, broad yellowish-colored tail, and a central light blue band or streak that runs the length of the dorsal and anal fins are additional helpful characters for identifying fresh-caught individuals.

Natural history notes.—*Caulolatilus princeps* has been reported at many localities from Willapa Bay, Washington, to the tip of Baja California, but they are quite rare north of Point Conception. They are very abundant throughout much of the Gulf of California, particularly in the northern third. They prefer living at depths from 30 to 450 feet or more where the bottom is rocky. They are much more abundant around offshore islands and on rocky banks than along the mainland coast.

Individuals that exceed 10 pounds are rarely seen, and 12 pounds and 40 inches probably are about maximum for weight and length. Six-year-old fish are 20 or 21

[46]

inches long, and weigh between 3 and 4 pounds. The oldest fish of several hundred large individuals examined was thirteen years old; this fish weighed 7½ pounds and was 25½ inches long.

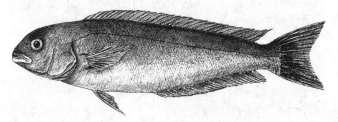

Fig. 13.   *Caulolatilus princeps*

Mature individuals, presumably ready to spawn, have been noted from October through April, but no information is available as to size and number of eggs spawned, age at first maturity, or various facets of reproductive behavior. Larval ocean whitefish have been captured in plankton nets many miles offshore, and young specimens just over an inch long have been found in the stomachs of albacore. These pelagic stages bear little resemblance to either young or adults in that they have numerous saw-toothed ridges on the sides and tops of their heads, plus a few other anomalous characters.

Ocean whitefish will feed upon a wide variety of fish, mollusks, and crustaceans that live in their preferred habitat. An examination of their stomachs has revealed shrimps, pelagic red crabs, hermit crabs, euphausiids, small octopi, squid, anchovies, blue lanternfish, and numerous other bite-sized creatures. Pelagic stages of *Caulolatilus* have been found in albacore stomachs, and adults are preyed upon by giant sea bass, sharks, and a few other large predators.

Otoliths of ocean whitefish were found in an Indian midden on San Clemente Island, and these fish were very likely important to the Indians who occupied most of the other Channel Islands.

[47]

Fishery information.—Ocean whitefish are not as highly prized by sport fishermen as they deserve to be. They put up an excellent battle when hooked, and those caught in fairly deep water (150 feet or deeper) are as delicately flavored as any food fish we have. Generally, 5,000 to 6,000 ocean whitefish are caught by partyboat fishermen each year, but during some years almost double this number has been taken.

The commercial catch is made exclusively with hook and line and is utilized by the fresh fish trade. Landings declined steadily from 1946, when 100,000 pounds were sold, through 1956 when less than a ton was delivered to the markets, and there has been little increase since 1956. This marked decline in landings reflects a lack of consumer demand that was brought about by an occasional bitter tasting whitefish. Market operators believe the bitter flesh was caused by food some whitefish had eaten. Many fish caught in and around kelp beds or in shallow water are so affected, but no ocean whitefish caught on one of the deep banks has ever had bitter flesh. The bitterness is like the taste of untreated tree-picked olives, and it cannot be removed by freezing or cooking, nor do affected fish that have been thoroughly cleaned immediately after they are caught taste any better than those left uncleaned until fishing has ceased for the day.

Other family members.—No other member of the family is known from California or closer to our shores than the Gulf of California.

Meaning of name.—*Caulolatilus* (differing from *Latilus*, a near relative, by the many fin rays) *princeps* (a leader).

Carangidae (Jack Family)
Yellowtail
Seriola dorsalis (Gill, 1863)

Distinguishing characters.—No other fish that inhabits the ocean off California has the body shape and color

[48]

of the yellowtail. Of eleven other jacks that have been reported from our waters, only two (Pacific amberjack and pilotfish) are similar enough to the yellowtail in body shape to cause possible confusion. The Pacific amberjack is a more compressed fish, has a rust-colored band that runs diagonally through the eye to the back near the dorsal fin, and has longer (almost falcate) anterior rays in the second dorsal and anal fins. The pilotfish has five or six vertical broad dark bands across its body.

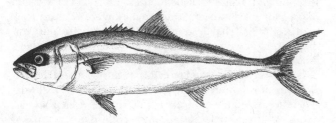

Fig. 14.   *Seriola dorsalis*

Natural history notes.—Seriola dorsalis ranges from Monterey Bay (during some years) to Cape San Lucas and is abundant throughout much of the Gulf of California. It is a schooling fish and usually is found close to shore, around islands and over offshore banks, but small schools are attracted to floating kelp and similar debris and may follow these objects offshore for many miles. The largest recorded yellowtail is an 80-pounder caught at Guadalupe Island, Mexico, a number of years ago. Some ichthyologists do not believe that our yellowtail can be distinguished from one in the Atlantic or from those off Japan, Australia, and Africa, so the 111-pound record listed by the International Game Fish Association in 1968 was a New Zealand fish. Several yellowtail caught at Guadalupe Island by a commercial fisherman in the late 1940s weighed between 90 and 115 pounds each, so it still may be possible for the yellowtail record to return to the eastern Pacific. A fish that weight is more than 5 feet long; one could only guess at its age. A

[49]

35-pound fish caught off California was twelve years old, but reliable ages are unavailable for larger individuals.

Many yellowtail will spawn when they are two years old and all will spawn when three. A three-year-old female will weigh about 10 pounds, and her ovaries will contain about 450,000 eggs; whereas, a 25-pound female will produce more than 1 million eggs. The spawning season normally runs from June through October, with July and August the peak months. During years when the ocean off southern California is warmer than normal, yellowtail have spawned successfully in our waters. Most spawning takes place off central and southern Baja California, however.

Yellowtail feed predominantly during the daytime and they are opportunistic feeders in that they will take whatever is edible and abundant in the area they inhabit. At times, pelagic red crabs comprise the bulk of the diet, while at other times anchovies are eaten in quantity. Squids, sardines, small mackerel, and other bite-sized items are also important in their diets. Yellowtail remains have been found in giant sea bass stomachs on a few occasions, and sea lions have been observed eating S. *dorsalis*, but man is unquestionably their greatest enemy, yet many die of old age.

Small groups of yellowtail have been observed nudging, bumping, and prodding live sharks (especially blue sharks) on many occasions. Since the shark being "pestered" has never been seen to turn on its "attackers," this observed byplay probably is another instance of parasite removal by a "cleaner fish" — in this case the yellowtail.

Fishery information.—Except for the flurry of excitement that a run of albacore causes among sport fishermen, yellowtail are the most sought after and highly prized game fish in the San Diego area and at most other southern California boat landings during some period or other each summer. For years the most productive sport fishing areas have been the waters around Santa Catalina and San Clemente islands, and along the mainland from south of Point Dume to Huntington Beach, between Dana Point and Oceanside, and off La Jolla and Point Loma. (The waters surrounding the Coronado Islands, Mexico, have been the best fishing grounds of all.) The annual California partyboat catch between 1936 and 1965 has been as low as 3,000 fish (1946) and as high as 457,000 (1959), but only during six of these years was it below 20,000 and only during three was it above 200,000.

The commercial catch has been made primarily in Mexican waters with purse-seine gear, but a fair poundage is taken off California each year also, although not with purse seines which are illegal for catching yellowtail in our waters. Most of the California catch is made with hook and line and in gill nets, and it is sold just as caught for the fresh fish trade. During the past decade the demand for canned yellowtail has been poor so the purse seine catch in Mexican waters has declined from around 10 million pounds in some years (1918 and 1936) to practically nothing in others. They make an excellent smoked product, but most yellowtail are consumed fresh (baked, broiled, fried, or raw).

Other family members.—Eleven other jacks have been taken off the coast of California, but five of these have been reported only once and there are fewer than ten records of occurrence for four others. Only S. *dorsalis* and *Trachurus symmetricus* are common components of our fauna, but *Decapterus hypodus* occurs with fair regularity during warm-water years.

Meaning of name.—*Seriola* (from the Italian vernacular for one member of the genus) *dorsalis* (pertaining to the long dorsal fin).

Jack Mackerel
Trachurus symmetricus (Ayres, 1855)

Distinguishing characters.—The sharp ridges on each side of the tail (modified lateral line scales), in conjunction with the long dorsal and anal fins and a deeply forked tail fin, will distinguish the jack mackerel from every other marine fish in our waters except the Mexican scad (*Decapterus hypodus*). The enlarged scales on the front part of the jack mackerel's lateral line (regular sized in Mexican scad), and the continuous dorsal and anal fins (the last dorsal and anal fin ray of *Decapterus* is separated as a finlet) will distinguish the jack mackerel from the scad.

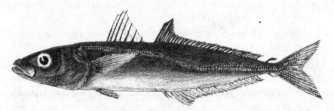

Fig. 15.   *Trachurus symmetricus*

Natural history notes.—Trachurus symmetricus has been captured between British Columbia and south of Magdalena Bay, Baja California, and ranges offshore for more than 500 miles. (A juvenile was picked up at Acapulco by a crew member on a tuna clipper, but there is a good possibility it had gotten there in the vessel's bait tank after having been netted somewhere around Magdalena Bay.) It is a pelagic schooling species, and most fishing for it is conducted around rocky headlands and offshore islands and banks.

Jack mackerel in the commercial catch usually are 8 to 15 inches long, but purse seiners sometimes catch 2½-footers with bluefin tuna, and sport fishermen seek out these giant-sized fish when they approach within range of the partyboat fleet. A 28½-inch fish weighed 5 pounds 4 ounces; the age of this fish was not determined, but others slightly larger have been over thirty years old.

Half of the two-year-old jack mackerel (10 inches to the fork of the tail) are sexually mature and will spawn

[52]

at that age; all are capable of spawning at age three. Spawning takes place from March through July over an extensive area 80 to 240 miles offshore. A single female will usually spawn more than once during this five-month period, and the tiny eggs (about one-twenty-fifth of an inch across) drift in the upper 300 feet of the ocean until they hatch in about four or five days.

Most of the food items found in jack mackerel stomachs have been euphausiids and amphipods, but at times *Trachurus* feeds heavily on pteropods, juvenile squid, small anchovies, lanternfish, and other handy organisms. Seals, sea lions, porpoises, swordfish, white seabass, giant sea bass, pelicans, and numerous other large fish-eating predators will capture and eat jack mackerel.

Otoliths and other identifiable remains of large *T. symmetricus* have been found in several coastal Indian middens. Their otoliths have also been found in many Pleistocene and Pliocene deposits of southern California, some estimated to be 10 or 12 million years old.

Fishery information.—When large jack mackerel are locally abundant they are much sought after by sport fishermen. The best fishing months are July through September, and the best bait is a large live anchovy. As many as 200,000 of these large jack mackerel have been caught by partyboat fishermen during a single season. Many others have been caught from barges, skiffs, and private cruisers, but there are no valid statistics on these catches.

The commercial catch was relatively insignificant until 1947 when the first attempts at large scale canning of jack mackerel were made. Since then, fish abundance and economics (consumer demand, warehouse inventories, prices paid fishermen, etc) have influenced the catch. As a result, annual landings have fluctuated widely reaching a low of 17 million pounds in 1954 and a peak of 146 million pounds in 1952.

Other family members.—Eleven other jacks have been captured off California at one time or another.

[53]

Most are instantly recognized as being jacks, but identification as to species is not always simple. For many it is necessary to count scales in the lateral line, as well as fin rays, and gill rakers; body shape, color, and fin lengths are also helpful characters.

Meaning of name.—Trachurus (rough tail) *symmetricus* (symmetrical). A symmetrical fish with a rough tail.

Clupeidae (Herring Family)
Pacific Sardine
Sardinops caeruleus (Girard, 1854)

Distinguishing characters.—The Pacific sardine is distinguished by a single small dorsal fin at about midlength, pelvic fins inserted at a point beneath the end of the dorsal fin, the lack of a lateral line, no scales on the head, a terminal mouth, and slightly raised ridges or striations running obliquely downward on the gill cover.

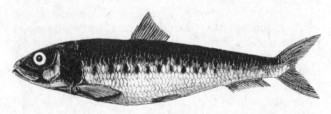

Fig. 16.   *Sardinops caeruleus*

Natural history notes.—*Sardinops caeruleus* has been captured at one time or another from southeastern Alaska to Cape San Lucas and throughout the Gulf of California. It is a schooling species and during periods of abundance off our shores some schools were believed to have contained 10 million or more individuals. A record fish was 16¼ inches to the fork of its tail, but those longer than 13 inches are veritable giants. Two very large individuals encountered recently were a ripe male 13½ inches long that weighed 15 ounces and a spawned-out female 14 inches long that weighed 13¼ ounces. Through the years, scales of tens of thousands of sardines have been used in age studies, and although not too reliable for old individuals, some fish were believed to have been twenty to twenty-five years old when captured.

Some fast-growing sardines will spawn when they are a year old, but most will not spawn until they are two. A one-year-old fish will average 5½ to 7½ inches in length — about half its expected total length. Spawning takes place at night, usually before midnight, and the eggs (each about 1.0 millimeter across) float or drift near the ocean's surface until they hatch in about three days. Each adult is believed to spawn two or three times during a season, extruding perhaps 90,000 to 200,-000 eggs annually.

Sardines typically are filter feeders, straining minute plants and animals from the sea as they swim, but at times they are selective feeders. In these instances they dart about and pick up certain kinds and sizes of animals to the exclusion of others. At various stages in their life cycle sardines are prey for just about every kind of flesh-eating animal in the sea. Copepods and arrow worms eat their eggs; jellyfish, anchovies and other creatures eat the larvae; and albacore, barracuda, yellowtail, sharks, pelicans, gulls, cormorants, sea lions, porpoises, and a multitude of other predators eat the juveniles and adults.

Sardine otoliths and identifiable fragmentia have been found in a few coastal Indian middens, but they are never abundant. There is no fossil record of *Sardinops caeruleus*.

Fishery information.—Large sardines can be caught on hook and line (Paula's lures and similar patented devices), and if the fisherman is using ultra-light tackle he is witness to a display of aerial acrobatics equal to that of the fightingest tarpon, but few individuals have indulged in the sport.

Commercially they have been captured with an assortment of roundhaul nets (purse seines, lamparas, bait nets, etc.) and through the years the bulk of the catch has been canned. Landings have ranged from "adequate to meet the demand," to "spectacular," to mediocre, to poor since the first California sardine cannery was established in 1889. Until 1915 annual landings fluctuated between 600,000 and 4 million pounds. They rose abruptly during the First World War and continued to increase to the 1.5 billion pound peak in 1936. The unprecedented decline since 1944 and eventual failure of the fishery is a well-known story. A moratorium declared in June 1967 may have come too late to be effective in rebuilding the population. On the other hand, there is some evidence (Indian middens, sediment cores, and lack of a fossil record) that *Sardinops* may never have been as dominant off California when no fishery existed as it was during the 1930s. The sardine might have "disappeared" whether it was fished for or not, because such catastrophic fluctuations have occurred naturally in many other fish groups.

Other family members.—Six other clupeids have been taken in the ocean off California at one time or another, but these can be readily identified by examining for a few salient features.

1. If there are striations on the gill cover, it is a Pacific sardine.
2. If the pelvic fins are attached well behind the dorsal, it is a round herring.
3. If the last dorsal ray is filamentous, and
   a. if the midline of the back in front of the dorsal fin is scaled, it is a thread herring.
   b. if the midline of the back in front of the dorsal fin is without scales, it is a threadfin shad.
4. If there are sharp, sawtoothed scales on breast and belly, and
   a. if the vertical rim on the edge of the shoulder girdle (beneath the gill cover) is smooth, it is an American shad.

[56]

b. if the vertical rim on the edge of the shoulder girdle has 2 fleshy lobes, it is a flatiron herring.

5. If none of the above major characters applies, it is a Pacific herring.

Meaning of name.—Sardinops (a fish with a sardine-like eye) *caeruleus* (blue).

Coryphaenidae (Dolphin Family)

Common Dolphin
Coryphaena hippurus Linnaeus, 1758

Distinguishing characters.—The elongate, laterally compressed, tapering body, which is highest at the point of insertion of the dorsal fin; the very long dorsal fin; and the profile of the forehead (vertical in adult males, rounded in females and young) are sufficient to distinguish the dolphin from other fishes in our waters. A live dolphin is a beautiful sight — "typically" it is brilliant turquoise blue or greenish above and silvery below, and has yellow pectoral and caudal fins. Upon being hauled from the water it changes color almost as if blushing, except that golden yellows predominate, among blues, greens, and other bright hues.

Fig. 17. *Coryphaena hippurus* (female)

Natural history notes.—Coryphaena hippurus has a worldwide distribution in warm seas, occurring sporadically off our coast during years of warm water, or during late summer months. When they are present in our area, they usually are in small schools (possibly ten to twenty-five or more individuals) and are hanging around floating patches of kelp, logs, or other flotsam in blue oceanic waters many miles from shore. They have been

[57]

caught by commercial albacore trollers as far north as Oregon. A record-sized dolphin caught in the Bahamas in 1964 weighed 76¾ pounds and was 5 feet 10½ inches long. There are reports of dolphins weighing in excess of 90 pounds. This is one fish species in which the male grows longer and weighs more than the female.

Their growth apparently is very rapid. A male that weighed just over 5 pounds when caught and put in an aquarium had added 43 pounds by the time it was accidentally killed less than a year later. Judged by this growth rate, maximum age for a dolphin would be about three years. Spawning in the eastern north Pacific apparently is in late spring and early summer, but this is based upon casual observation and not upon a definitive study. We have no information on age at first spawning (possibly one year), or on number of eggs spawned per season.

The dolphin stomachs we have examined have contained mostly small surface-dwelling fishes of many kinds, including small jacks, triggerfish, porcupinefish, flyingfish, and mackerels to name a few. We have also found fair numbers of cephalopods (mostly argonauts and squids) and crabs. Small dolphins probably are preyed upon by many species, but we have found them only in tuna and jack stomachs. Large dolphins have been reported in marlin and shark stomachs.

Fishery information.—When dolphins are present in our waters, it pays sport fishermen to investigate every floating patch of kelp and debris they encounter. A handful of anchovies cast next to the flotsam will bring all kinds of action if fish are there; baited hooks are the next order of business. If the angler does not desire to stop, trolling an artificial feather, metal lure, or plastic jig past the floating object is equally as effective. The best catch by partyboat fishermen was made in 1957, a warm-water year, when a reported 3,000 dolphins were taken. The catch from skiffs and private boats is not known.

There is no commercial fishery for dolphins in our waters, which is difficult to understand since the flesh is highly prized and thousands of pounds are flown and shipped into California for use in restaurants. The menu listing for this top-notch food and game fish is *mahi-mahi*.

Other family members.—No other member of the family is known from north of about Magdalena Bay.

Meaning of name.—*Coryphaena* (in allusion to the males having [or showing] a helmeted head) *hippurus* (horse tail).

Cottidae (Sculpin Family)
Cabezon
Scorpaenichthys marmoratus Girard, 1854

Distinguishing characters.—The cabezon is one of forty-two sculpins known to inhabit ocean waters off California. It is readily identified by the smooth, scaleless, wrinkled-appearing skin, and a prominent cirrus on the snout. Its stout body, broad head, large mouth, and a long frilly cirrus at the upper rear margin of each eye are additional helpful characters.

Fig. 18.   *Scorpaenichthys marmoratus*

Natural history notes.—*Scorpaenichthys marmoratus* ranges from Sitka, Alaska, to Abreojos Point, Baja California. The larvae and young are pelagic, but when about 2 inches long, they move inshore and take up residence in rocky habitat from the intertidal into depths exceeding 250 feet.

[59]

This is another species with a larger reported weight (25 pounds) than can be authenticated, although a 39-inch fish caught in May 1962 at Point Lobos State Reserve may have exceeded the reported maximum weight. Unfortunately, this veritable giant was not weighed. A 28½-inch fish weighed just over 15 pounds and was thirteen years old, judged by growth zones on its otoliths.

Males first mature when they are two years old (13½ inches long), but females do not mature until they are a year older (17½ inches long). As with many other fish species, females grow faster, attain larger sizes, and live longer than do males.

Spawning takes place from November through March, usually at "community nesting sites" where adults congregate year after year. The eggs are laid in large masses on rocks that have been cleared of growths and debris by the fish, and individual nests are guarded by the males until the eggs have hatched. A 3-pound female will produce about 50,000 eggs, while a 10-pounder will produce nearly 100,000.

Cabezons feed heavily on crabs and mollusks (abalones, squid, clam siphons, octopi, and others), but they will also capture and eat small fish. Sea lions are known to eat an occasional cabezon, and they would not be difficult for a sea otter to catch, but we do not know of any specific predator on large adults.

Otoliths of S. *marmoratus* have been found in several Pleistocene deposits in southern California, and they are common in some Indian middens on the coast north of Point Conception.

Fishery information.—Rocky-shore anglers often fish specifically for cabezon, but most of the catch is made by fishermen who are after anything that will take a baited hook. Of an estimated 38,000 taken annually by all sport fisheries north of Point Arguello, 28,000 are caught by shore anglers, 5,000 are caught from skiffs, just over 1,000 are speared by skindivers, and the rest are taken from piers and partyboats. An additional 14,-000 are taken each year in southern California, mostly from shore.

Cabezon have never been an important commercial species. A peak catch of 34,000 pounds was landed in 1952, but during most years poundages have averaged less than one-third of this amount. Most of the commercial catch is made with setlines incidental to rockcod fishing. They are marketed fresh, but since the flesh is often bluish-green in color (although it becomes white when cooked), there is little demand for them. Cabezon roe has been proven toxic and can make a person violently ill if eaten, so the roe should always be discarded.

Other family members.—For information on the forty-two species of marine cottids known from California please refer to our checklist (p. 161) and to our list of references (p. 172).

Meaning of name.—*Scorpaenichthys* (*Scorpaena* fish, for its resemblance to the scorpionfish) *marmoratus* (marbled).

Echeneidae (Remora Family)
Common Remora
Remora remora (Linnaeus, 1758)

Distinguishing characters.—The first dorsal fin of the remora is modified into an adhesive disk that is crossed by a series of laminae, which represent the fin spines. The disk of the common remora has seventeen to twenty laminae, and there are twenty-one to twenty-seven dorsal fin rays, and twenty-six to twenty-nine pectoral rays, which distinguish it from all other members

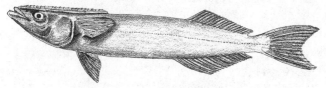

Fig. 19.   *Remora remora*

[61]

of the family. It usually is necessary to remove one side of the leathery sheath which covers the dorsal fin in order to count the underlying rays.

Natural history notes.—Remora remora ranges throughout all tropic and temperate world seas. Juveniles to perhaps 3 inches in length are free-living, but at progressively larger sizes they are associated with progressively larger sharks to which they adhere with their dorsal disk. They can release themselves at will, and usually do so to pick up or catch food, or when their host has been caught and is being hauled from the water.

Two individuals that were caught on hook and line off California in the 1950s appear to be as large as the species gets. They were both 32 inches long, and one of them weighed 14 pounds. The fisherman who caught them had been fishing for albacore and did not observe any sharks in the area, although there usually are large blue sharks around a boat where albacore are being caught. The remoras, however, may have been free-living.

We have no information on any facets of reproduction, and data on food habits are scanty. One of the most recent publications on remora behavior indicates they are predators, symbionts, and commensals. Large individuals are most apt to have fish remains in their stomachs, which they have probably captured themselves. Smaller sizes feed heavily on parasitic crustaceans (sea lice) that they remove from the body of their host, but in addition they often eat scraps that are dropped by their host while it is feeding. Off California, the host shark usually is a blue, but during warm-water years when hammerheads are present common remoras often associate with these, and they have been observed in the company of mantas.

Remoras have been found in the stomachs of sharks, marlins, tunas, and sea birds. They undoubtedly are eaten by most other large predators that live in the same environment.

[62]

Other family members.—Five other members of the family have been found off California at one time or another. These can be identified by counting disk laminae, dorsal rays, and pectoral rays. In some cases, the host or body color is sufficient to distinguish the species.

1. If there are 9 to 11 disk laminae, it is a slender suckerfish.
2. If there are 12 to 13 disk laminae it is a white suckerfish.
3. If there are 14 to 20 disk laminae, and
   a. 21 to 27 dorsal rays and 20 to 23 pectoral rays, it is a marlinsucker.
   b. 21 to 27 dorsal rays and 26 to 29 pectoral rays it is a common remora.
   c. 28 to 33 dorsal rays, it is a spearfish remora.
4. If there are 21 to 28 disk laminae, it is a whalesucker.

Meaning of name.—*Remora* (an ancient name for the "holding back" ability of the fish) *remora* (repetition of generic name).

Embiotocidae (Surfperch Family)
Barred Surfperch
Amphistichus argenteus Agassiz, 1854

Distinguishing characters.—The barred surfperch is one of three surfperches living off our sandy beaches whose sides have bronze, brassy, or yellow-gold bars and spots against a whitish background. It can be distinguished from the other two (calico and redtail) by its lower jaw being slightly shorter than the upper, and by its spiny dorsal fin being shorter than the rays in the soft dorsal. The calico and redtail surfperches also have reddish-colored tails and fins, a color never found in the barred.

Natural history notes.—*Amphistichus argenteus* ranges from Bodega Bay, California, to Playa Maria Bay, Baja California. It is most abundant in the breaking surf along sandy beaches, but it sometimes ranges into deeper water offshore and there is a record of one being caught in 240 feet of water. The largest individual known was a 17-inch female that weighed 4¼ pounds and was nine years old. The oldest male of 984 individuals examined during a recent study was six years old; this fish was 12 inches long. Five of 1,142 females examined during the same study were seven to nine years old and ranged in length to 16 inches. Thus, as with many other fish species, female barred surfperch grow both larger and older than males.

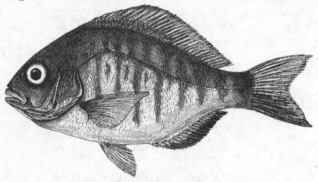

Fig. 20 *Amphistichus argenteus*

Barred surfperch, as do all members of the family, give birth to living young. Mating usually takes place in the fall and early winter and the young are born from March to July. At birth, these youngsters are junior replicas of their parents, and are independent in that they swim off on their own and must fend for themselves. Adult females shorter than 10 inches long average about 25 fish per "litter," while those larger than 10 inches average about 45; a record litter consisted of 113 young.

Ninety percent of the barred surfperch diet is sand crabs, with bean clams and small crustaceans making up the bulk of the rest of their food. Small barred surfperch are eaten by several predators that live or feed in their environment (including birds), but large fish have few enemies other than man. They apparently live out their lives within a few miles of where they are born.

[64]

PLATE 1.

1. Grunion digging in as wave breaks on shore.

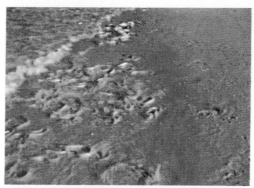

2. Spawning grunion left on beach by receding wave.

3. Female grunion attended by two males.

4. Spawned-out grunion waiting for wave to inundate the area and carry them back to sea.

PLATE 2.

1. A mixed school of half-moon and small jack mackerel.

2. Small school of jack mackerel.

3. Schoollike aggregation of salema.

PLATE 3.

1. Typical aggregation of rubberlip perch.

2. Black perch in bed of giant kelp.

3. Juvenile sheephead staying close to low growing algae over rocky substrate.

4. Giant sea bass swimming a few feet above sandy bottom. Polka dots disappear when fish is removed from water.

PLATE 4.

1. Diver's view of moray and several bluebanded gobies.

2. A sculpin in a typical pose with venomous dorsal spines erect.

3. Cabezon braced against surge.

4. Male kelp greenling.

5. Female kelp greenling.

PLATE 5.

1. Kelp bass, sheephead, garibaldis, and señoritas attracted to a free meal of sea urchins broken up by divers.

2. Señorita removing parasites from blacksmiths at an underwater "cleaning station."

PLATE 6.
1. Kelp rockcod.

2. Vermilion rockcod.

3. Gopher rockcod.

4. Black-and-yellow rockcod.

5. Flag rockcod (front) and starry rockcod (rear).

6. Brown rockcod.

Plate 7.

1. A string of rockcod on a partyboat.
2. Rockcod fishing reels attached to rail of partyboat. These large reels ease the task of pulling up a string of fish from 600 to 800 feet.
3. Rockcod fishing with conventional rod and reel.

Plate 8.
1. Aerial photo of a purse seiner setting its net around a school of bonito (dark area in center of net).

2. Pole fishing for yellowfin tuna — a technique made obsolete by the modern purse seine fleet.

3. A typical catch made with an otter trawl in shallow water— mostly stingrays, shovelnose guitarfish, and flatfish.

Out of 1,987 fish tagged, 209 were recovered, and few had traveled more than 2 miles during the period they were at liberty. The greatest movement was 31 miles, recorded for one adventuresome individual.

Otoliths from *A. argenteus* have been identified from Pliocene deposits near San Diego and San Pedro, and barred surfperch remains are abundant in many coastal Indian middens. The fossil otoliths may be 12 million years old.

Fishery information.—South of Point Arguello, the barred surfperch has been reserved exclusively for sport fishermen since 1953. Some are caught by pier fishermen but the bulk of the catch is made by shore fishermen on sandy, waveswept beaches. Successful baits include sand crabs, bloodworms, salt- or sugur-cured mackerel, mussels, clams, artificial lures (especially small spinners and large wet flies or streamers). Barred surfperch often make up 70 to 80 percent of the sandy-shore fisherman's bag in southern California, with an estimated 114,000 being taken annually from 1963 through 1965, the only years for which such estimates are available. North of Point Arguello an estimated 273,000 were caught per year by sport fishermen between 1958 and 1961 — 51,000 from piers and 222,000 from shore.

Few barred surfperch are taken by commercial fishermen even though they are abundant along some beaches north of Point Arguello where commercial fishing is permitted. Unfortunately, the commercial catch is reported only as "perch" or "saltwater perch" and at least eight species are involved, so no poundage estimate is available for barred surfperch.

Other family members.—Twenty surfperches inhabit our waters, nineteen in salt water and one in fresh. The various species are distinguished by a wide variety of characters including body and fin color, thickness of lips, lengths of fin spines and rays, number of lateral-line scales, number of dorsal and anal fin rays, and so forth.

Meaning of name.—Amphistichus (double series) *argenteus* (silvery).

Rubberlip Seaperch
Rhacochilus toxotes Agassiz, 1854

Distinguishing characters.—The thick fleshy lips (white or pinkish in color) are sufficient for identifying a rubberlip seaperch. The spines in the dorsal fin are all shorter than the soft-rayed portion of the fin, and there are twenty-seven to thirty rays in the anal fin.

Natural history notes.—Rhacochilus toxotes ranges from Little River, Mendocino County, California, to Cape Colnett, Baja California. It is found in quiet waters of harbors and bays, and is often abundant in kelp beds, around piers and jetties, and outside the surf along the open coast. It is said to reach a length of 18 inches and probably is the largest member of the family, but no studies have been made on the fish, so there is but scanty information on many facets of its life history.

R. toxotes is viviparous. A 16½-inch female speared in the Redondo Beach area during June contained 21 young that averaged just over 3½ inches each. The parent weighed slightly less than 3 pounds, and was eight years old. Rubberlip seaperch with nearly mature embryos have been taken during April through June, so the

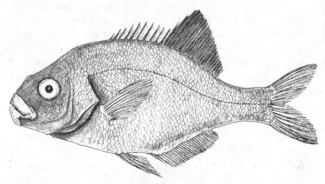

Fig. 21.   *Rhacochilus toxotes*

[66]

spawning season probably is nearly identical to that of the barred surfperch.

The few stomachs that have been examined have contained crustaceans almost exclusively, including shrimp, amphipods, small crabs, and small stomatopods.

Fishery information.—Rubberlip seaperch are taken by pier and jetty fishermen, skiff fishermen, shore fishermen, and skindivers. Most hook-and-line catches are made with mussels, clams, sand worms, cut shrimp, and similar bait. They tend to travel in small aggregations or schools (10 to 25 or more individuals), so fishing in a given spot often will yield several nice "keepers" in a short period. From 1958 through 1961, sport fishermen caught an estimated 5,000 rubberlip seaperch per year in the area between Point Arguello and the Oregon border. The annual catch south of Point Arguello is believed to average about double this amount.

This is one of the eight to ten kinds of perch that are important in the commercial fishery. In the Monterey area it usually is one of the two leading species of perch caught with gill nets, but it is less important elsewhere. In 1967, the statewide "perch" catch was only 201,000 pounds; rubberlip seaperch probably contributed less than 10 percent of this total.

Other family members.—For information on other family members please refer to p. 65, to the checklist of game and food fishes (p. 162) and to the list of references (p. 172).

Meaning of name.—*Rhacochilus* (rag lip) *toxotes* (in allusion to the East Indian archer fish because of some presumed resemblance).

Engraulidae (Anchovy Family)
Northern Anchovy
Engraulis mordax Girard, 1854

Distinguishing characters.—The overhanging snout, single dorsal fin at mid-body, lack of a lateral line, and very large mouth and gill opening will distinguish northern anchovies from all non-anchovies in our waters. The gill covers that are not united ventrally separates them from the anchoveta, and the head that is longer than the body is deep separates them from the other three anchovies.

Natural history notes.—Engraulis mordax ranges from British Columbia to about Cape San Lucas. Young fish form dense schools and usually are found closer to shore than older individuals. Large mature adults often (though not always) form smaller schools and tend to remain in offshore waters where they go fairly deep (400 to 600 feet) during daylight hours, but they move to the surface and disperse at night. Northern anchovies are said to attain a maximum length of 9 inches, but very seldom is one seen that exceeds 7. Most species of *Engraulis* do not live longer than two or three years, but many *E. mordax* reach four years and some attain seven. An 8-inch fish, the largest we have seen, weighed 2 ounces; it was six years old.

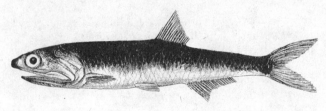

Fig. 22.   *Engraulis mordax*

Female anchovies often spawn when they are 4½ to 5 inches long and about a year old. Some spawning takes place during every month of the year, and a single fish will spawn more than once during a given season. In all, from 20,000 to 30,000 eggs will be released by each female during a spawning season. These oval-shaped eggs drift with the currents at or near the surface of the ocean and hatching occurs in about three days.

Anchovies are filter feeders, straining micro- and macro-plankton from the water as they swim. At times they will eat fish eggs and larval fish, perhaps even their

[68]

own relatives, but mostly they feed on crustaceans and other tiny organisms. Anchovies, however, are fed upon by just about every fish-eater that swims in or flies over the ocean. Porpoises and swordfish can and will eat a bucketful of anchovies at a time, whereas a white croaker might eat only two or three. Unfortunately (for the anchovy, that is), there are more white croakers in our waters than swordfish and porpoises. Fortunately (for the predators and mankind) there are enough anchovies to go around.

Otoliths of *E. mordax* have been found in goodly numbers in every coastal Indian midden that has been searched for them, and in every Pleistocene and Pliocene exposure between Mexico and Eureka that has been examined for fish remains. Some of the Pliocene deposits are believed to be more than 10 million years old.

Fishery information.—There is no sport fishery for northern anchovies, but thousands of tons are netted each year for use as live bait by partyboat and other fishermen. Many additional tons are salted for use as dead bait, or ground for chum.

From 1948 until the mid 1960s an increasing quantity of the commercial anchovy catch was canned for both human consumption and pet food. In 1966, legislation was passed making it legal to catch a limited tonnage of anchovies for processing into fish meal and oil (reduction), and the commercial catch increased accordingly. Researchers had determined that an estimated 4 million tons of anchovies were living off our coast and it was felt desirable to harvest the unused surplus. Those empowered to control and manipulate the quota (California Fish and Game Commission) proceeded cautiously, so the allowable tonnage during the first few years of this reduction fishery was not all the commercial fishermen had wished for, but it was more than the state's sport fishermen wanted because they still had a vivid memory of what commercial overexploitation had done (or was thought to have done) to the once plentiful sardine.

Other family members.—Four other anchovies have been recorded from California's waters, but one of these (slim anchovy) has been noted only once, while one other (anchoveta) is rare. The five anchovies are relatively easy to identify.

1. If the gill covers are joined to each other in the throat region, it is an anchoveta.
2. If the body is nearly round in cross section, the head is longer than the body is deep, and the color is dusky blue above, it is a northern anchovy.
3. If the body is laterally compressed, the head is shorter than the body is deep, the color is whitish above, and
   a. there are 17 to 20 anal rays, it is a slim anchovy.
   b. there are 23 to 26 anal rays, it is a slough anchovy.
   c. there are 29 to 33 anal rays, it is a deepbody anchovy.

Meaning of name.—*Engraulis* (ancient name for the common European anchovy) *mordax* (biting — in allusion to the large mouth).

Exocoetidae (Flyingfish Family)
California Flyingfish
Cypselurus californicus (Cooper, 1863)

Distinguishing characters.—The evenly-pigmented winglike pectoral and pelvic fins are sufficient to identify the California flyingfish. Additional characters include an unbranched first pectoral ray, a single dorsal fin placed far back on the body, a tail fin with the lower lobe longer than the upper, and a normal-appearing lower jaw (not produced or beaklike).

Natural history notes.—*Cypselurus californicus* has been recorded from Astoria, Oregon, to Cape San Lucas, Baja California, but it seldom is observed north of Point Conception, and knowledge of its distribution south of Cedros Island is vague. During daylight hours, adults are scattered at the surface over a vast expanse of the ocean, but at night they are known to move toward island shores where they are believed to feed; before

[70]

dawn they again head out to sea. This behavior has contributed to a successful gillnet fishery for flyingfish at
Santa Catalina and San Clemente islands.

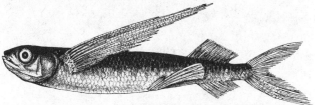

Fig. 23.   *Cypselurus californicus*

The largest of several hundred flyingfish examined
during recent years was a female 19 inches long that
weighed about 1¼ pounds. A fish that was 1¼ inches
shorter weighed 2 ounces more, but the enlarged ovaries of this shorter fish weighed nearly 3 ounces. Individuals as large as these two appear to be five years old, if
reliance can be placed upon the growth zones noted on
their otoliths.

Spawning takes place at the surface of the ocean
from late June into September, and the eggs usually adhere to floating debris. Eggs of *C. californicus* average
1.64 millimeters in diameter (0.0656 of an inch), and
have approximately 60 long filaments attached uniformly over the egg surface. They hatch in about sixteen days. No information is available regarding growth
rates of the young fish. The largest adults are always
females.

Most flyingfish captured in gill nets at night, or in
purse seines, have had no food in their stomachs, but
among those containing food, small crustaceans and larval fish have been the dominant items noted. Tapeworms sometimes are found in the intestines of flyingfish, but generally they are free of observable parasitic
worms. Flyingfish are heavily preyed upon by sea birds,
especially boobies, and by porpoises, and dolphins. Marlins, tunas, and other large surface-feeding species also
eat flyingfish when the opportunity arises, and two large

[71]

flyingfish were found in the stomach of a 12-pound moray speared at San Clemente Island.

Remains of *C. californicus* have not been found in any of California's fossil deposits.

Fishery information.—Commercial fishermen take advantage of the diurnal migrations that *C. californicus* makes and catch them during summer months in gill nets strategically placed around Santa Catalina and San Clemente islands. The fish caught in these nets, which are set parallel to the shoreline just before dark, are almost invariably headed inshore when snared. Catch records have been kept only since 1928, but during most years, landings have held fairly steady at around 30,000 to 60,000 pounds per year. The peak catch was 171,000 pounds in 1965. Most of the catch is frozen or iced and used as bait by marlin fishermen in various oceans of the world. Flyingfish flesh is oily and not as mildly flavored as dozens of other fish, but it is highly esteemed as food by many people.

Other family members.—Two other flyingfishes have been reported from off California, but the otoliths and other characters of one of these (*Fodiator acutus*, the sharpchin flyingfish) indicate that it belongs in the halfbeak family. Many ichthyologists believe the halfbeaks and flyingfishes belong to a single large family, but their otoliths are sufficiently distinctive to retain them in separate families. We have included *Fodiator* in our key (below), but not as a member of Exocoetidae in our checklist.

1. If the lower jaw is produced into a bony mental process, it is a sharpchin flyingfish.
2. If all but the first and second pectoral rays are branched, it is a blackwing flyingfish.
3. If all but the first pectoral rays are branched, and
   a. if there are fewer than 45 (usually 40 to 42) predorsal scales, and the posterior margins of the pectorals are blackish, it is a blotchwing flyingfish.
   b. if there are more than 45 (usually 48 to 50) predorsal scales, and the pectoral fins are evenly dusky above, it is a California flyingfish.

[72]

Meaning of name.—Cypselurus (based upon the name of the European swift which it resembles in "flight") *californicus* (from California).

Gadidae (Cod Family)
Pacific Tomcod
Microgadus proximus (Girard, 1854)

*Distinguishing characters.—*The three dorsal fins and two anal fins distinguish members of the cod family. The Pacific tomcod can always be identified by a combination of three characters: a lower jaw that is shorter than the upper, a chinwhisker that is shorter than the eye diameter, and the presence of the anal opening vertically beneath the posterior part of the first dorsal fin.

Fig. 24.   *Microgadus proximus*

Natural history notes.—Microgadus proximus ranges from Alaska to Point Sal, California, and is the commonest member of the family found in California's waters. It is especially abundant in depths of 100 to 300 feet where the bottom is either sand or firm sandy mud, but at times large numbers are found just outside the surf zone in only a few feet of water.

It is reported to attain a length of "about a foot," and it probably does, but there are no actual measurements for an individual that large. The largest we have examined was a 10½-inch female that weighed 6 ounces and was two years old. We have examined a fair number of 8- to 9-inch individuals and these were all one year old. The maximum age probably does not exceed five years. We have no information on maturity, fecundity, or re-

[73]

production, but have noted ripening eggs in two-year-old fish.

Pacific tomcod feed heavily upon shrimp but will eat other bottom-dwelling crustaceans as well as clams and small fish. In turn, they are fed upon at all sizes and ages by a wide variety of predators. Their remains have been found in the stomachs of several kinds of flatfish, and in salmon, hake, rockcod, lingcod, sea lions, and porpoises, to name a few.

Otoliths of *M. proximus* are abundant in several Pleistocene and Pliocene deposits that outcrop in northern California and Oregon. Some of these deposits were laid down about 12 million years ago.

Fishery information.—In central and northern California, Pacific tomcod are caught primarily by shore and pier fishermen when inshore runs occur. Skiff fishermen also catch a few, but they are generally ignored by partyboat fishermen and skindivers. An estimated 8,000 were being caught per year during 1958 to 1961, according to figures obtained during a survey of sport fisheries north of Point Arguello.

Most of the commercial catch is made with shrimp and otter trawls, but the species is not highly prized for food so annual landings seldom exceed a ton or two. Some of the landings are utilized in the fresh fish trade, but most are processed for pet food or used by fur farms.

Other family members.—Two other members of the family are taken off our shores; both of these attain lengths of 3 feet and will weigh several pounds each when full grown. The three cods are relatively easy to identify by observing a few simple characters.

1. If the lower jaw is longer than the upper it is a walleye pollock.
2. If the lower jaw is shorter than the upper jaw, and if:
   a. the anus is under the front part of the second dorsal fin and the chinwhisker is longer than the eye it is a Pacific cod.

[74]

b. The anus is under the back of the first dorsal fin and the chinwhisker is shorter than the eye it is a Pacific tomcod.

Meaning of name.—Microgadus (small *Gadus*) *proximus* (near, in reference to its close affinity with *Microgadus tomcod* an Atlantic species).

Hexagrammidae (Greenling Family)
Kelp Greenling
Hexagrammos decagrammus (Pallas, 1810)

Distinguishing characters.—The presence of five lateral lines on each side is characteristic of the genus *Hexagrammos.* Within this genus, the kelp greenling is distinguishable from all other fishes by its color. Males have numerous turquoise-blue blotches on the head and front part of the body, and each of these is ringed by small rust-colored spots. Females are uniformly covered with round reddish-brown spots.

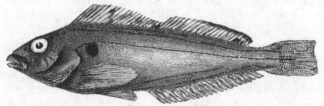

Fig. 25. *Hexagrammos decagrammus*

Natural history notes.—Hexagrammos decagrammus has been recorded from Kodiak Island, Alaska, to La Jolla, California, but south of Point Conception it lives in deeper water and is much less common. Throughout their range they live in rocky habitat, especially if there is a good cover of algae. In cold waters to the north, they are common intertidally, and are fairly abundant into depths of 75 feet. Off southern California, they have been hooked in 150 feet of water.

A 21-inch kelp greenling speared at a skindiving meet appears to be a record size, but this fish was not weighed nor were other vital statistics taken. A 16-inch

[75]

female weighed just over 2 pounds, and its otoliths indicated it was eleven or twelve years old. It appeared to have spawned for the first time when it was three years old.

Spawning occurs in the fall and early winter, and the mass of pale blue eggs adheres to the rocky substrate. We have no information on number of eggs per female, time required to hatch, nesting behavior (if such occurs), or other facets of reproduction.

The kelp greenling feeds upon an assortment of small crustaceans, polychaete worms, small fishes, clam siphons, and other edible organisms that live in rocky habitats. In turn, their remains have been found in stomachs of lingcod, sharks, and a couple of kinds of rockcod. Ospreys have been observed catching greenlings in intertidal areas, and sea lions are believed to eat them occasionally.

Fishery information.—The kelp greenling is one of the most important species in the rocky-shore fisherman's bag along the north coast. They will take almost any type of bait offered, including cut fish, clams, mussels, shrimp, squid, worms, and small crabs. Shore anglers generally catch much smaller fish than do skindivers, or skiff and partyboat fishermen. More than 56,000 are taken each year by California's sport fishermen, according to results of a survey conducted from 1958 to 1961.

They are not a sought-after commercial species, and fewer than 1,000 pounds are taken incidental to other fisheries. These are absorbed by the fresh fish trade, usually in the form of fillets.

Other family members.—Although some modern ichthyologists place the members of three other families (Ophiodontidae, Oxylebiidae, and Zaniolepididae) into family Hexagrammidae, we prefer to retain the combfishes in their own family (Zaniolepididae). Thus, we recognize only five Californian species as belonging to the expanded greenling family. These can be distinguished by observing an assortment of characters.

[76]

1. If there is but 1 lateral line and:
   a. there are large hooked canines in the jaws (body with gold spots), it is a lingcod.
   b. the jawteeth are hardly noticeable (body with dusky red vertical bars), it is a convictfish.
2. If there are 5 lateral lines and:
   a. if the dorsal fin is continuous (unnotched), it is an Atka mackerel.
   b. if there is a deep notch between the spinous and soft dorsal fins and:
      i. the fourth lateral line fails to reach as far as the pectoral fins, it is a whitespotted greenling, *Hexagrammos stelleri*. (This species has been erroneously reported from California, and is included here only on the chance that it may stray into our waters at some future date.)
      ii. the fourth lateral line reaches far back on the body, the fifth lateral line divides at about the midlength of the ventral fins, and the tail fin is rounded, it is a rock greenling.
      iii. the fourth lateral line reaches far back on the body, the fifth lateral line divides opposite the last third of the ventral fin, and the tail fin is slightly notched, it is a kelp greenling.

Meaning of name.—Hexagrammos (six-line — the fifth lateral line on each side divides) *decagrammus* (ten-line, for the total of ten lateral lines on both sides).

Lingcod
Ophiodon elongatus Girard, 1854

Distinguishing characters.—The general body shape, long dorsal fin containing more than twenty-four spines and twenty soft rays, lower jaw that projects beyond the upper, and large mouth filled with long sharp caninelike teeth are sufficient to distinguish a lingcod. The color is extremely variable but usually ranges from dark grayish-blue to greenish-brown, and there are numerous dark blotches and mottlings on the sides. Numerous golden spots on the sides are very characteristic of fresh-caught adults.

Natural history notes.—*Ophiodon elongatus* ranges from northwestern Alaska to the vicinity of San Martin

[77]

Island, Baja California. Young fish often are found in bays and other quiet-water areas where the bottom is sandy, but adults live at or near the bottom in close association with a rocky substrate. They are most abundant at depths shallower than 350 feet, but south of Point Conception few large lingcod are caught in less than 500 feet of water. They have been taken at depths greater than 1,000 feet.

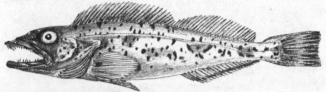

Fig. 26.   *Ophiodon elongatus*

A 5-foot long 70-pounder was reported from British Columbia in 1866, but we believe these are not actual measurements, but estimates. The largest authentic length and weight record for California appears to be a 45-inch female caught off Monterey that weighed 41½ pounds. There undoubtedly are larger lingcod, but we do not have supporting data. A 40-pound fish may be sixteen to twenty years old.

Male and female lingcod first mature when they are three years old. At that time they will be about 26 inches long and will weigh about 4 pounds each. Spawning takes place from December through March, and depending upon size, a female will produce from 60,000 to 500,000 eggs per spawning (30- and 45-inch fish). The eggs are large (1/6 inch in diameter) and adhesive. They stick in large masses to the rocky substrate and the male lingcod remains on guard over the nest until the eggs hatch.

Tagging studies have shown that adult lingcod usually do not travel or migrate any great distance. One fish was recovered at almost the exact spot it had been tagged twelve years earlier.

[78]

Young lingcod depend heavily upon shrimp and other crustaceans for sustenance, but adults rarely eat anything except fish and an occasional octopus or squid. They will take any size fish they can get their mouth around. Lingcod have been eaten by the white shark, and we suspect that they are preyed upon by sea lions.

Lingcod remains have been found in several Indian middens along the central California coast.

Fishery information.—North of Point Conception the lingcod sport fishery is very important the year around, but this species is not heavily fished in southern California except during winter months when partyboats travel to offshore banks in search of rockcod. The state's partyboat fishery has averaged over 30,000 lingcod per year since 1947 (ranging from 13,000 in 1953 to nearly 45,-000 in 1958), and an estimated 25,000 more have been taken annually by other sport fisheries, including skindivers.

The commercial catch is made primarily with multi-hook setlines and with otter trawls. Landings have varied between about 400,000 and 2 million pounds annually since 1916, with the peak catch in 1948. Catches have been below 1 million pounds during most recent years.

Lingcod are utilized almost exclusively in a fresh state, primarily in the form of steaks and fillets. The flesh of some individuals is a greenish color but is in no way harmful, and all traces of green disappear with cooking.

Other family members.—For information on the six members of the family, please refer to p. 77.

Meaning of name.—*Ophiodon* (snake tooth) *elongatus* (elongate).

Icosteidae (Ragfish Family)
Ragfish
Icosteus aenigmaticus Lockington, 1880

Distinguishing characters.—The general shape of the spotted young ragfish cannot be mistaken for any other marine species in our area, especially when its soft flabbiness is taken into consideration. The adult is brownish (without spots), thich-skinned, oblong-ellipti-cal, and has a slightly forked tail, but the very flabby body and troutlike head are still quite apparent. Pelvic fins are missing in the adult.

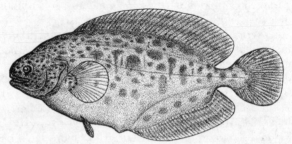

Fig. 27.   *Icosteus aenigmaticus* (juvenile)

Natural   history   notes.—Icosteus   aenigmaticus ranges from northern Alaska to San Onofre, California, and across the north Pacific to Japan. Juveniles are spot-ted (as figured) and supposedly inhabit great depths, but fair numbers have been captured in relatively shallow water near the shore or near the surface offshore. Adults apparently do live in deep water, if reliance can be placed upon capture methods and gear. The largest measured individual was 6 feet 10 inches long, but larger specimens have been taken. A 5-foot 3-inch fe-male caught off Eureka in July 1966 weighed 54 pounds.

An examination of otoliths of several ragfish 10 inches to 5 feet long indicates that the spotted phase lasts less than a year. When these fish grow to be 10 to 12 inches long they probably move offshore to areas that are sel-dom fished, for not many are captured until they are 2 feet long or longer. A few individuals 2 to 2½ feet long have been caught near the surface in salmon gill nets about halfway across the north Pacific. Growth is quite rapid during the first two years, and spawning probably

[80]

occurs in the third or fourth summer. Most of the 4- to 5-foot adults captured off California are females nearly ready to spawn. These seven- to nine-year old fish apparently enter our coastal fishing grounds for spawning purposes.

The ovaries of five large individuals (three captured during July and one each during November and January) contained three sizes of eggs. The largest eggs were slightly smaller than 1/8 inch in diameter, and between 230,000 and 430,000 were present in each fish. All of these fish appeared to be in spawning condition, so some spawning obviously occurs during both summer and winter.

The only individuals examined for food had eaten small fishes, squids, and octopuses. Large ragfish have been found in the stomachs of several sperm whales.

Fishery information.—Spotted juveniles show up frequently in trawl catches, but gill nets and purse seines also have yielded fair numbers. There is at least one record of a juvenile ragfish being caught on hook and line. The Japanese occasionally catch adult ragfish in gill nets, and although one adult has been reported as taken on hook and line, all the large fish we have seen were caught in trawl nets being fished on the bottom in depths of 45 to 200 or more fathoms (270 to 1,200 feet). One or two large ragfish have been cast ashore by stormy seas.

Other family members.—*I. aenigmaticus* is the only member of this family known.

Meaning of name.—*Icosteus* (literally, a fish with yielding bones, alluding to its flabby raglike consistency) *aenigmaticus* (puzzling) — a puzzling fish with a flabby body.

Istiophoridae (Billfish Family)
Striped Marlin
Tetrapturus audax (Philippi, 1887)

Distinguishing characters.—The rounded, rather elongate bill or spear, movable pectoral fins, and "triangular" dorsal fin will distinguish the striped marlin from all other fishes in our waters.

Natural history notes.—Tetrapturus audax appears only seasonally (summer and fall) off southern California, ranging north to about Point Conception during some years. It is found throughout much of the subtropics and tropics to Chile, and around Hawaii, Fiji, the Marquesas, Korea, Japan, the Philippines, New Zealand, Formosa, and many other Pacific localities. A 465-pound fish that was 10 feet 6 inches long was the International Game Fish Association's world record in 1968; it was caught off New Zealand in 1948. The 692-pounder caught at Santa Catalina Island many years ago and long recognized as the world record striped marlin was almost certainly a black marlin. Very few marlin taken off California will exceed 250 pounds. Nothing is known about marlin ages but a 220-pounder probably is somewhere between three and five years old.

Striped marlin larvae were not recognized until 1959 when a Japanese biologist described a growth series from localities in the North Pacific, South Pacific, and Indian oceans. Spawning in the North Pacific apparently takes place between May and August, but no spawning grounds are known within several thousand miles of California. A 154-pound striped marlin landed in Honolulu had one ovary weighing 32.2 pounds that was estimated to contain nearly 14 million eggs.

Marlin feed heavily upon pelagic species of fish (sauries, sardines, jack mackerel, and flyingfish in our area) and squid, and crabs, but they are also known to feed upon mesopelagic species several hundred feet beneath the surface. Marlin have few enemies aside from man, although many are infested with a copepod parasite, *Pennella* sp., and others will have peculiar stalked barnacles attached to their skin. Frequently, marlin will be caught that have one or more remoras adhering to them, either in plain sight on the body or under a gill cover with just the tip of the tail showing. Sometimes more than one species of remora is present.

[82]

We do not know of any fossil record of *Tetrapturus* from California, nor have their remains been found in Indian middens as yet.

Fig. 28.　*Tetrapturus audax*

Fishery information.—It has been illegal to take marlin commercially off California since 1937, so only a sport fishery exists at present. Almost the entire sport catch is made by trolling live or dead bait, or some type of artificial lure, behind a small boat in known marlin waters. In early years, flyingfish were thought to be the only bait for marlin, but since the Second World War large sardines (when available), or medium-sized jack mackerel, and Pacific mackerel have proven highly successful. The best fishing locality is a triangle of water bordered by Santa Catalina and San Clemente islands on the north and extending south to the Coronado Islands.

Fresh marlin is delicious whether smoked, baked, or broiled, so long as it is not allowed to cook dry. The Japanese value the flesh for sashimi (raw).

Other family members.—Three other species are found off California but none of the three has been captured (reported) more than three or four times. Billfishes, distinguished from swordfish by their rounded spear, are easily identified by a few simple characters.

1. If the pectoral fins are rigid it is a black marlin.
2. If the entire dorsal fin is very high it is a sailfish.
3. If the bill is barely longer than the lower jaw it is a shortbill spearfish.
4. If none of the above characters apply it is a striped marlin.

[83]

Meaning of name.—Tetrapturus (four-wing tail, in reference to the winglike caudal keels) *audax* (bold or rash).

Kyphosidae (Sea Chub Family)
Opaleye
Girella nigricans (Ayres, 1860)

*Distinguishing characters.—*The bright blue eyes, and olive-green perchlike body will distinguish the opaleye from all other species in our waters. Most individuals have one or two white spots on each side of the back near the middle of the dorsal fin.

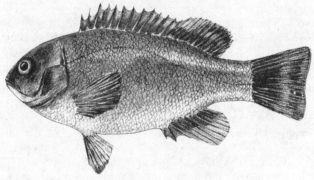

Fig. 29. *Girella nigricans*

Natural history notes.—Girella nigricans ranges from San Francisco (where it has been reported once) to Cape San Lucas, although it is rare south of Magdalena Bay. It is a year-around resident south of Point Conception but is seen only rarely north of there. The earliest stages are silvery colored and lead a pelagic existence. At about an inch they show up in small aggregations in the rocky intertidal region which they call home until they are about 3 inches long. Larger sizes and adults are found near the bottom in rocky subtidal areas, particularly where there are kelp beds. They are most abundant in depths of about 20 feet, but are plentiful in almost any rocky habitat in 5 to 65 feet of water.

[84]

The largest opaleye we have seen was a 25 3/8-inch long fish speared near Laguna Beach in 1964 that weighed 13½ pounds. Growth rings on its otoliths indicated it was ten or eleven years old. Females that were ready to spawn have been taken in April, May, and June, which probably corresponds to their main spawning season. We have no information on the number of eggs spawned by a single female per season, time required for the eggs to hatch, or other facets of reproduction. They apparently mature and spawn for the first time when they are about 8 or 9 inches long and two or three years old.

Stomachs and intestines of opaleyes usually are packed with algae: fleshy and coralline reds, as well as browns and greens. Eelgrass has also been found in their stomachs. Apparently the animals that live on the seaweeds are necessary to their diet, whereas the plant life itself supplies little or no energy to the opaleye. Individuals fed a diet of clean seaweed (with all animal life removed) have failed to gain weight and have died in about a month. We know of no specific predator of adult opaleye.

Fishery information.—The opaleye is one of the most important species in the rocky-shore fisherman's bag south of Point Conception, and the second most important species to spearfishermen during competitive meets. Because of their small mouth and the cautious way they take a bait, few fish are more difficult to hook. Once hooked, their ability to do battle combined with the rocky habitat in which they live make them even more difficult to land. An assortment of baits will appeal to a hungry opaleye, particulary green sea moss, green peas, mussels, and sand crabs. Southern California sport anglers were taking an estimated 74,000 opaleyes per year according to a survey conducted during 1963-1965.

The commercial catch is made almost exclusively with encircling nets incidental to other fisheries, although a power-operated lift net was used with considerable success shortly after the Second World War. Since 1953 the commercial catch has varied between about 1½ to 5 tons per year. The entire quantity is sold as "perch" in the fresh fish markets.

Other family members.—We have followed one of the most recent proposals for classification at family and higher levels in placing the opaleye in the sea club family. Under this system, two other fishes are family relatives: the halfmoon and the zebraperch. These three laterally-compressed (perchshaped) fish can be distinguished by a few easily observed features.

1. If the fish is olive green and has bright blue eyes, it is an opaleye.
2. If the fish is an overall blue-gray (darker above) and the soft dorsal and anal fins are covered with a thick sheath of scales, it is a halfmoon.
3. If the body is silver-gray or yellow-gray and there are about twelve black vertical bars on each side, it is a zebraperch.

Meaning of name.—*Girella* (from the French *Girelle*, a name applied to small labrids) *nigricans* (blackish).

Labridae (Wrasse Family)
California Sheephead
Pimelometopon pulchrum (Ayres, 1854)

Distinguishing characters.—The three California members of this family are easily distinguished by their "buck" teeth which are forward projecting and canine-like. The sheephead can be told from the other two wrasses by its body color regardless of size, age, or sex. Young fish are a solid orange-red and have seven round black blotches, five of which show in side view on the fins and at the base of the tail. Adult females are a dull red or rose color, and adult males have bluish-black heads and tails and a red midsection. Both sexes have white chins.

Natural history notes.—*Pimelometopon pulchrum* ranges from Monterey Bay to Cape San Lucus and throughout much of the Gulf of California. They are

[86]

found near the bottom in rocky areas, particularly where kelp forests abound. They are most abundant at depths of 10 to 100 feet, but are found both shallower and deeper. Because of pollution, disappearance of kelp beds, fishing pressure (including spearing by skindivers), and other unfavorable factors, they are no longer a common inhabitant of rocky areas along our mainland coast, although small to medium-sized individuals still abound around the Channel Islands. Large males (exceeding 20 pounds) are rare off our coast.

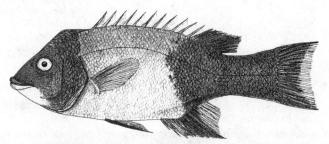

Fig. 30. *Pimelometopon pulchrum* (male)

A male sheephead speared off Point Loma in November 1956 was reported to weigh 36¼ pounds, and probably is a record size. The largest we have seen was a 29-pound male that was 32 inches long. This fish was fifty-three years old, so it is easy to see why heavy fishing pressure would make drastic inroads on large fish. The largest female we know of, an 18¼-pound fish, was thirty years old. One individual is known to have lived in an aquarium for twenty-three years before dying.

Casual observations indicate that sheephead probably first spawn when they are four or five years old. Most spawning takes place during summer, and the eggs are believed to be free-floating. The sheephead probably changes sex (from female to male) at some stage of its life, but proof of this is lacking.

Underwater observations and examination of several dozen stomachs have revealed that sheephead feed on a

[87]

wide variety of fixed or slow-moving creatures that live in their environment. Such items as sea urchins, sand dollars, mussels, kelp scallops, small snails, abalones, lobsters, hermit crabs, tube-dwelling polychaetes, squids, and octopi have been noted with fair regularity. Remains of medium-sized sheephead have been found in the stomachs of giant sea bass, but man appears to be their worst enemy.

Fossilized jaw and pharyngeal teeth of sheephead have been found in many Pliocene and Pleistocene deposits of Southern California, and jaw teeth that appear to be from *P. pulchrum* have been recovered from several Miocene exposures. These may be 20 to 25 million years old. Sheephead remains, especially teeth, are abundant in coastal and insular Indian middens.

Fishery information.—Sport fishermen seldom consider a sheephead as being a prize catch, yet southern California partyboat fishermen have consistently taken about 15,000 per year during the past two decades, and the annual catch reached a record 53,000 in 1966. The catch from skiffs, private boats, and the rocky shore is unknown, but probably equals or exceeds the partyboat catch. Skindivers, particularly during competitive meets, expend extra effort to spear large sheephead in their quest for trophies which are awarded for total fish poundage. Sheephead flesh is fine-grained, white, and mild in flavor; it makes excellent table fare whether fried, baked, broiled, used as stock in chowder, or to complement lobster or crab meat in a salad.

Between 1926 and 1950, commercial landings often exceeded 100,000 pounds per year, with a peak of 370,-000 pounds in 1928. Since 1950, however, landings seldom have exceeded 20,000 pounds, and have been as low as 5,000 to 7,000 pounds during some years. The specialized nature of the fishery (hook and line) and a lack of consumer demand have been the primary contributing factors for these poor landings.

Other family members.—The three wrasses that inhabit our waters are easy to distinguish by making a few simple observations.

1. If the fish has 12 spines in the dorsal fin it is a California sheephead.
2. If it is cigar-shaped, orange-brown in color and has a large black blotch at the base of the tail it is a señorita.
3. If neither of the above applies it is a rock wrasse.

Meaning of name.—*Pimelometopon* (fat forehead) *pulchrum* (beautiful). A beautiful fish with a fat forehead.

Lampridae (Opah Family)
Opah
Lampris regius (Bonnaterre, 1788)

Distinguishing characters.—The iridescent disk-shaped body with white spots and crimson fins distinguishes the opah (sometimes called moonfish) from other fish.

Fig. 31.   *Lampris regius*

[89]

Natural history notes.—In the eastern Pacific, specimens of *Lampris regius* have been caught far north in Alaska and near Cape San Lucas, Baja California, and at many intermediate localities. Offshore they range across the Pacific, and are also found in the Atlantic Ocean, both north and south of the equator. Although opahs have been reported to attain weights of 500 to 600 pounds and lengths of 6 feet, we have not found any authentic record of a weight of even 200 pounds. The heaviest of several hundred individuals from our coast weighed 160 pounds and was 4½ feet long.

Opahs apparently will eat whatever food is available in their area and often in great quantities. The stomach of a 126-pounder caught near Anacapa Island in 1959 contained the remains of 63 fishes (51 hake, 4 rockcod, 1 brotulid, and 7 unidentifiables), 8 cephalopods, and 7 pelagic crabs. Another opah, caught on longline gear several hundred miles offshore, had eaten 5 lancetfish, each 3 to 4 feet long.

We have no information on ages, spawning habits, or migrations, although a large female caught in the early spring appeared to be nearly ready to spawn.

Fishery information.—During albacore season each year a few fishermen catch opahs off California, especially in the seas between the northern Channel Islands (Anacapa, Santa Cruz, Santa Rosa, and San Miguel) and the Coronado Islands. Some of these fish are hooked near the surface and some quite deep, some on live bait and some on trolled artificial lures. Opahs are taken in fair abundance in tuna longline fisheries conducted on the high seas. These longline-caught fish are usually hooked 400 to 1,000 feet beneath the surface. Occasionally an opah becomes entangled in a gill net which has been set for other species. The salmon-colored flesh of the opah is quite tasty, but rather dry when cooked; it is best when smoked.

Other family members.—*Lampris regius* is the only living member of the family known at present, although several other scientific names have been applied to opahs throughout the world. A Miocene fossil opah found near El Capitan (Santa Babara County) was named *Lampris zatima*.

Meaning of name.—Lampris (brilliant) *regius* (king).

Luvaridae (Louvar Family)
Louvar
Luvarus imperialis Rafinesque, 1810

Distinguishing characters.—The tunalike outline, dolphinfishlike head, crimson fins, and frothy pink body (silver when dead) distinguish the juvenile and adult louvar from all other fishes. Its bones are cartilaginous or weakly ossified, and break, tear, or pull apart when any strain is placed upon them.

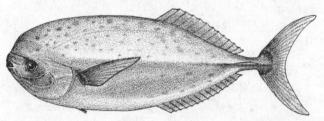

Fig. 32. *Luvarus imperialis*

Natural history notes.—On our coast, specimens of *Luvarus imperialis* have been reported from Acapulco, Mexico, and Newport, Oregon, but most of the fifty or more known individuals have been from the area between San Diego and Monterey. Louvars inhabit most other world seas, but do not seem to be as abundant elsewhere as off California. Our purse-seine fisheries probably account for the record of their abundance in our waters.

If such records were kept in the world, California apparently would hold them for both the longest louvar (a 6-foot 2-inch fish from off Huntington Beach in 1955) and the heaviest (a 305-pounder, washed ashore at Redondo Beach in 1932). Louvar otoliths are very tiny and do not provide any indication of age.

The louvar feeds primarily upon jellyfishes and similar gelatinous planktonic forms. In order to derive maximum food value from these, the digestive tract is modified in several ways: the stomach is lined with many elongate, nipplelike projections, which increase the absorptive surface, and the intestine is very long and convoluted. A 40-inch fish, which weighed 45 pounds, had a 37-foot intestine. Only a few individuals examined have had any food in their stomachs, and this has consisted of jellyfish, ctenophores (comb jellies), pyrosomes, and one small fish.

The ovaries of a 295-pound female that washed ashore at Morro Bay in May 1953 were enlarged and contained nearly ripe eggs. This would indicate late spring or early summer spawning in our area. Nothing is known about growth rates or ages, but an 8-inch louvar found in the stomach of a wahoo at Clarion Island, Mexico, in April 1953 could have been from the previous year's hatch.

Fishery information.—Most of our local louvars have been caught at night by purse-seine vessels fishing for bluefin tuna, especially near San Clemente Island. Many have been cast ashore where they were found by beach strollers or fishermen; others have been picked up at sea where they were drifting dead at the surface or floundering in a feeble condition. A couple have been reported as having been hooked, but such reports are extremely doubtful because of the weakly constructed jaws and other bones. One louvar was caught in a shark net, one was killed by an underwater explosion, and one was found in a wahoo stomach. The louvar has very delicate white flesh and is considered tasty by those privileged to try some.

Other family members.—*L. imperialis* is the only known member of the family, but because of the metamorphosis its odd-shaped larva goes through, and its world distribution, many other scientific names have been given to it.

Meaning of name.—Luvarus (silver?) *imperialis* (emperor).

MERLUCCIIDAE (HAKE FAMILY)
Pacific Hake
Merluccius productus (Ayres, 1855)

Distinguishing characters.—The typical fin positions and shapes of the Pacific hake, combined with such characters as a large mouth filled with numerous sharp teeth, very deciduous scales, slightly forked tail fin, and protruding lower jaw which lacks a barbel will preclude mistaking it for any other fish in our waters with the possible exception of the walleye pollock. Its soft flabby body, dusky silvery coloration on back, and pectoral fin which reaches past the end of the first dorsal are more than sufficient to distinguish it from the walleye pollock.

Fig. 33.   *Merluccius productus*

Natural history notes.—Merluccius productus ranges from northwestern Alaska to Magdalena Bay, and offshore for at least 350 miles. There are reports of hake having been captured at depths of nearly 3,000 feet (500 fathoms), but the great majority of those off California appear to live shallower than 750 feet. They are a schooling species, and apparently undertake vertical migrations as well as horizontal — moving primarily inshore and off, seasonally. They are known to attain a length of 3 feet and a weight of 8 to 10 pounds, but a 30-inch fish is rare; the largest we were able to measure

[93]

was 26 inches long and weighed 4 pounds. A fish this size would be about fifteen years old. As with many other species, females grow faster and larger than males.

Both males and females are mature when they are 16 inches long and four years old. Spawning is primarily in the winter, and each fish is said to spawn only once during a season, in spite of there being two distinct sizes of eggs in the ovaries just prior to spawning. Hake eggs taken in plankton tows will average 1.07 to 1.18 millimeters in diameter (.0428 to .0472 of an inch). Ovaries of a 15-inch hake will contain about 50,000 eggs, while those of a fish twice that length will contain perhaps ten times that many.

Stomachs of juvenile hake have been gorged with shrimplike crustaceans and small squid primarily, but some small fish are also present. Adult hake feed mostly upon other fish and apparently do not discriminate as to what kind they eat — taking whatever kind is handiest when they are hungry. They also eat fair quantities of shrimp and squid as well as an occasional clam. In turn, hake are a favorite prey for a great many creatures, especially marine mammals (seals, sea lions, porpoises, small whales, etc.). Adult hake have also been found in the stomachs of swordfish, lingcod, soupfin sharks, Pacific halibut, electric rays, and an assortment of other large, carnivorous fishes.

Otoliths of *M. productus* have been found in several coastal Indian middens — presumably the remains from meals the Indians had eaten a few hundred to several thousand years ago. Otoliths have also been recovered from numerous Pleistocene and Pliocene deposits, some estimated as being 10 to 12 million years old.

Fishery information.—There is no sport fishery for hake, although they would be easy to catch in unlimited quantities much of the time. They are not desirable as food because of their soft flabbiness, tastelessness, and poor keeping quality. If cleaned and cooked within min-

[94]

utes of being caught, however, they are said to rival the best of the food fishes for flavor and general edibility.

Until the Russians became interested in our hake resources, the bulk of the commercial catch, made almost exclusively with trawling gear, was sold for animal food. Catches made by the Russians (many thousands of tons annually since 1967) have been processed by their factory ships for human consumption within the U.S.S.R.

Other family members.—No other member of the family is known on the outer coast north of Cape San Lucas. *M. angustimanus*, a species that is found in the northern Gulf of California, may be a variant of *M. productus*.

Meaning of name.—*Merluccius* (the ancient name meaning sea pike) *productus* (drawn out).

Molidae (Mola Family)
Ocean Sunfish
Mola mola (Linnaeus, 1758)

Distinguishing characters.—Once an ocean sunfish has been seen it could not be mistaken for any other fish. The oval silver-colored body which is lacking in a tail fin; high dorsal and anal fins set well back on the body; tiny terminal mouth; thick layer of cartilage beneath the skin; and lack of pelvic fins are characteristic of this fish.

Natural history notes.—*Mola mola* is found in all temperate and tropical seas of the world. In the eastern north Pacific, ocean sunfish have been noted from Alaska to well below the Mexican border. They are sporadically abundant off California, particularly from the San Francisco area south.

A record ocean sunfish that was struck by a Russian ship in the Tasman Sea in 1965 was reported to be 4 meters long (13.1 feet) by 2 meters high (6.56 feet) and to weigh more than 1,500 kilograms (3,300 pounds). Because these measurements and weights are in round figures, they obviously are estimates. The next largest

[95]

appears to be a 10-foot 1-inch male that was 11 feet from the tip of the dorsal fin to the tip of the anal fin. A 1965 report of a large *Mola* from off Argentina gives its weight as 2,640 pounds, considerably less than the 1,410 kilograms (3,000 pounds) reported for one in 1903 (this latter weight undoubtedly is an estimate and very likely is erroneous). Few individuals from off California will exceed 3 feet in length or 125 pounds. These small ocean sunfish are most commonly observed at the surface where they lie on one side and lazily flap their fins. Sometimes these fish will be seen flipping several feet into the air and dropping back into the water with a loud splash.

Fig. 34.   *Mola mola*

Their spawning grounds are not known, and fertile eggs and early larvae have not been found, although a 4-foot 6-inch female was estimated to contain 300 million eggs. Tiny post-larvae, some as small as 1/4 inch, have been captured in plankton nets in the open ocean on several occasions, but even these are rare. These tiny pea-shaped fish are covered with "a cuirass of broad plates with jutting triangular projections." These spiny plates disappear with growth.

Most stomachs examined have been empty, but jellyfish, ctenophores, crustaceans, small fish, and pelagic mollusks have been found fairly often. Ocean sunfish are heavily infested with tapeworms and other parasites, which undoubtedly cause some mortality. Fishermen often report that gulls and other sea birds will be "pecking" at ocean sunfish as they lie on their sides at the surface. Possibly these birds are removing sea lice or other external parasites. Surfperches and señoritas have both been observed removing parasites from ocean sunfish at nearshore "cleaning stations" in Monterey Bay and other coastal areas. Aside from large sharks and sea lions, ocean sunfish probably have few enemies.

Fishery information.—Ocean sunfish are not the object of a fishery, either sport or commercial, and they are not believed edible by most fishermen although some find them a delicacy. They are often brought ashore by skindivers, skiff, and partyboat fishermen as curiosities; these fish usually have been speared, gaffed, or dipnetted while they were "basking" at the surface, but several individuals have been caught with hook and line during the past two decades. Purse-seine fishermen usually catch them incidentally while netting anchovies, mackerel, or sardines, or if a crewmember wants some to eat he will use a gaff or dipnet on "basking" fish.

Other family members.—One other member of the family, *Ranzania laevis*, has been recorded from California on two occasions. The two species are easily distinguished by their color, body shape and size, and by their mouths.
1. If the body is twice as long as deep and the lips (mouth) cannot be closed it is a slender mola.
2. If body is about one and a half times as long as deep and the lips can be closed it is an ocean sunfish.

Meaning of name.—Mola (mill wheel, for its resemblance to a circular milling stone) *mola* (a repetition of the generic name).

Mugilidae (Mullet Family)
Striped Mullet
Mugil cephalus Linnaeus, 1758

Distinguishing characters.—The striped mullet is easily distinguished from all other fishes by its two widely separated dorsal fins, forked tail, anal fin with eight or nine soft rays, and small terminal v-shaped mouth. Atherinids (jacksmelt and relatives) might be mistaken for mullet the first time one sees them, but they have twenty-two or twenty-three soft rays in the anal fin.

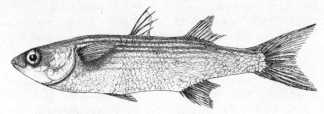

Fig. 35.   *Mugil cephalus*

Natural history notes.—Mugil cephalus has been reported as ranging from about Monterey Bay, California, to Chile, but in recent years it seldom has been seen north of Playa del Rey. At times mullet have been very abundant in Salton Sea where they have provided flourishing commercial and sport fisheries. They are most commonly observed in coastal estuaries, bays, lagoons, and flowing waterways (irrigation canals and such) throughout the Imperial Valley and from Los Angeles County south. They aggregate in small schools comprised of a dozen to several hundred individuals, and if they are present in a given body of water, a few individuals will be seen jumping at the surface at almost any time of day. They are reported to attain lengths of 3

[98]

feet, but the largest noted in Salton Sea during investigations of the fishery were 2 feet long and weighed just under 10 pounds. Many of these large fish were sixteen years old, which is about maximum for the species.

M. cephalus apparently spawns during winter months in the eastern Pacific, and it is generally believed that spawning takes place well offshore over deep water. Silvery post-larvae and juveniles are often captured in plankton nets and with dip nets near the surface many miles at sea. At just over an inch in length they commence moving shoreward and 2- to 5-inch individuals are abundant in shallow water. Mullet up to a foot long or longer commonly enter freshwater streams and rivers and some individuals have been known to travel fifty miles or more upstream. Female mullet attain larger sizes than males.

The pelagic larvae are eaten by numerous surface feeding predators including tropical tunas, whereas juveniles to 6 inches or so are preyed upon rather heavily by fish-eating birds (pelicans, gulls, terns, frigate birds, and such). Mullet food habits have been studied in many parts of the world, and diatoms, blue-green algae, green filamentous algae, cladocerans, and other microscopic plants and animals comprise the bulk of their diet.

Mullet otoliths have been found in 25 million-year-old Miocene deposits near Bakersfield, but these do not appear to be from *M. cephalus.*

Fishery information.—Mullet seldom will take a baited hook, but in irrigation canals and other freshwater habitat where they are abundant, they sometimes can be caught with a tiny hook that has been baited with a small fragment of earthworm, or with a tiny dry fly. When mullet were abundant in Salton Sea fish were commonly snagged with a surf-casting outfit and a large treble hook while they aggregated in shallow water near the mouths of various freshwater inlets. In conjunction with this method, clubs, spears, and dipnets were also used effectively.

Commercial fishing has been almost entirely with gill nets, primarily in Salton Sea and the salt ponds in south San Diego Bay. The fishery has been sporadic at both localities because of variations in mullet abundance, gear restrictions, closed seasons, poor market conditions, and other factors. In the past fifty years the commercial catch has ranged from about 1,300 pounds in 1941 to a peak of 503,000 pounds in 1945. The flesh is white but quite oily; it makes a topnotch smoked product.

*Other family members.—*No other member of the family is known from California, although at least seven other mullets exist in the tropical eastern Pacific.

Meaning of name.—Mugil (to suck, in reference to the mullet's presumed feeding behavior) *cephalus* (pertaining to the head, which in the mullet is quite distinctive).

Muraenidae (Moray Family)
California Moray
Gymnothorax mordax (Ayres, 1859)

*Distinguishing characters.—*The California moray lacks pectoral and pelvic fins, has a leathery skin, has low fleshy ridges where the dorsal and anal fins would be, and its jaws are filled with well-developed, serrate, sharply pointed teeth.

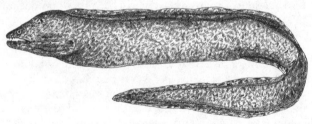

Fig. 36   *Gymnothorax mordax*

Natural history notes.—Gymnothorax mordax ranges from north of Santa Barbara to Santa Maria Bay, Baja California, and it is especially common around the shores of our offshore islands. It is a secretive species during daylight hours but can be found in almost any rocky area to depths of about 130 feet if the right kinds of hiding places are examined. They are reported to attain a length of 5 feet, which they unquestionably do, but the largest we have measured was a 47 1/2-inch long male that weighed 14 1/2 pounds. The age of this fish is not known, but a California moray lived in an aquarium for twenty-six years.

Their favorite food appears to be shrimp and crabs, but they will come out of hiding and tear into abalones that skindivers injure in loosening them from rocks. The 14 1/2-pounder mentioned above had eaten two fair-sized California flyingfish.

No information is available regarding their spawning season, egg-laying ability, age at maturity, larval stages, or in fact, where they are and what they look like up to a length of about 8 inches. Lobster fishermen often catch morays in their traps, and if only one or two are present in a trap with a dozen or more lobsters, the morays usually are dead when the trap is pulled. If about equal numbers of morays and lobsters are entrapped, however, both will still be alive.

Moray teeth have been identified in coastal Indian middens and a Pleistocene deposit estimated to be nearly 2 million years old.

Fishery information.—Shore fishing at night in a rocky habitat, especially at Santa Catalina Island, will yield excellent catches of morays. Cut squid, shrimp, anchovy, or abalone make excellent bait, but once hooked the moray must not be given any slack or it will end up wedged into a rocky hole or crevice and the angler will get a broken line for his efforts. Another technique that yields good catches is to poke-pole for them at low tide. For this, the fisherman uses a bamboo pole about 10 feet long with a wire leader about 6 inches long attached to the small end and a large hook attached to the wire. The end of the pole with its baited hook is inserted into likely crevices, holes, and other hiding places. When a moray is hooked it is unceremo-

[101]

niously hauled from its hole and subdued with a baseball bat before being unhooked and sacked. There is no commercial fishery or sale for morays; those taken in lobster traps usually are dumped back into the sea.

In preparing a moray for the table, some individuals peel the tough hide off with the aid of a sharp knife, but parboiling the beheaded and eviscerated moray greatly simplifies the task of exposing the mild white flesh.

Other family members.—No other member of the family is known within several hundred miles of California.

Meaning of name.—*Gymnothorax* (naked breast in allusion to the lack of scales) *mordax* (prone to bite).

Osmeridae (Smelt Family)
Night Smelt
Spirinchus starksi (Fisk, 1913)
Distinguishing characters.—The white body, abdominal pelvic fins, small adipose fin, relatively large mouth (maxillary extends to near back edge of eye), lack of striations on the gill cover, nine to thirteen gill rakers on the upper limb of the first gill arch, and five to eight pyloric caeca will distinguish the night smelt from all other fishes inhabiting the waters of our coast.

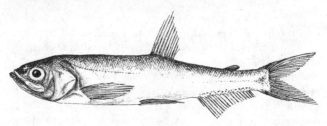

Fig. 37.   *Spririnchus starksi*

Natural history notes.—*Spirinchus starksi* ranges from Shelikof Bay, southeast Alaska, to Point Arguello, California. They spawn in the surf at night from January through September and prefer beaches where the

[102]

sand is coarse and gravelly, so they are not at peak abundance south of Monterey. There are no records of night smelt exceeding 6 inches in length, and no valid information is available regarding their longevity but it is believed that some individuals reach an age of at least three years. During spawning, the males congregate in dense schools in quite shallow water and the females dash in, extrude their eggs and return to deeper water. At the height of runs males outnumber females by about 8 to 1 and at the end of the runs they are present in a ratio of about 100 to 1. The fertilized eggs sink to the bottom in the surf zone where they stick to the sand grains and are eventually covered by several inches of sand. Hatching takes about two weeks but no information is known regarding larval and early juvenile stages. Adults presumably spawn more than once during the season.

Their food consists of small shrimplike crustaceans. Night smelt are preyed upon by just about every fish-eating predator in their habitat: birds, seals, and a wide assortment of fish.

Otoliths of *S. starksi* have been found in an Indian midden south of Morro Bay, and they are common in several Pleistocene deposits (300,000 to 500,000 years old) that have been investigated between San Pedro and Eureka.

Fishery information.—An occasional night smelt is caught by a pier fisherman using tiny hooks and a small piece of bait, but the bulk of the sport and commercial catch is made with two-man jump nets or one-man A-nets. In using the two-man net, the fishermen jump an incoming wave, plant the end poles (with the net stretched between them) firmly on the bottom in an upright position, and the fish are caught in the net as the wave recedes. The A-net, invented and used by North Coast Indians long before the white man showed up, is fished in a similar fashion but can be used to catch either incoming or outgoing schools. A good haul will yield 25 to 50 pounds of fish. Beaches most famous for runs of night smelt are Moss Landing, Scott Creek, Martins Beach, Halfmoon Bay, Portuguese Beach, Russian River, Juan Creek, Centerville Beach, Gold Bluff, Mad River Beach, and Smith River Beach.

Commercial fishermen catch as much as 200,000 pounds of smelt (mostly night and surf smelt) each year and sport fishermen take many additional pounds, but commercial catch figures are not separated by species and no survey has been made of the sport fishery, so exact figures are not available. Since large oceanariums became popular, much of the night smelt catch is sold to them for porpoise and sea-lion food. Night smelt make excellent table fare, needing only to be eviscerated and descaled before crisp-frying them, either with or without a coating of batter.

Other family members.—Five other osmerids are known from our coastal waters, and these are not easy to distinguish at a glance, so an examination of internal anatomy may be necessary until one becomes familiar with all the family members.

1. If the opercle (gill cover) is striated, it is a eulachon.
2. If the maxillary (upper jaw) fails to reach to beneath the center of the eye (when the mouth is closed), and
   a. If the head is four times as long as the eye or longer, and there are 66 to 73 rows of scales along the side of the fish, it is a surf smelt.
   b. If the head is less than four times as long as the eye (about 3 1/2 times), and there are 53 to 58 scales along the side, it is a delta smelt.
3. If the maxillary reaches past the rear margin of the eye, and
   a. if the eye diameter exceeds four-fifths the depth of the caudal peduncle, it is a whitebait smelt.
   b. if the eye diameter is less than four-fifths the caudal peduncle depth, and
      i. the pelvic fin is long, reaching past the anus, it is a longfin smelt.
      ii. the pelvic fin is much shorter, failing to reach the anus, it is a night smelt.

[104]

Meaning of name.—Spirinchus (pointed snout) *starksi* (for Edwin C. Starks, 1867-1932, noted West Coast ichthyologist).

Pleuronectidae (Righteye Flounder Family)
Starry Flounder
Platichthys stellatus (Pallas, 1814)

*Distinguishing characters.—*The starry flounder is the easiest to identify of all our flatfishes. The dorsal, anal, and caudal fins have black bands and streaks, and the body is covered with raised plates or stellate scales that are rough and rasplike to feel. Although they are members of the righteye flounder family, about 60 percent have their eyes on the left side.

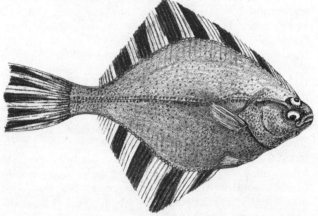

Fig. 38.   *Platichthys stellatus*

Natural history notes.—Platichthys stellatus is distributed throughout the north Pacific Ocean, both east and west, having been recorded south to Santa Barbara on our coast, but seldom seen south of about Morro Bay. They live over mud, sand, or gravel bottom areas, and are most abundant in shallow coastal waters including bays, sloughs, and estuaries, but they also invade streams and rivers and have been caught offshore in depths of 1,000 feet.

[105]

They are reported to reach a length of 3 feet and a weight of 20 pounds, but these are such round figures that one must be suspicious of them, especially since there does not appear to be a positive first-hand account of such a large fish. During a recent survey of fisheries north of Point Arguello, a 15-pounder was reportedly examined, but its vital statistics are not available. A 25 1/2-inch female weighed about 7 1/4 pounds and was seven years old, as was a 6-pound female that was 23 1/4 inches long. We have no information on maximum age, but believe it could exceed fifteen years; a large individual from the Bering Sea was thirteen.

Males spawn at the end of their second year when they are about 14 1/2 inches long, but females seldom spawn until age three when they are about 16 1/2 inches long. The usual spawning season extends from November through February, with peak activity in December and January. An estimated 11 million eggs are spawned by a 25-inch female during a single season. Hatching takes place about two and one-half to three days after the eggs are fertilized, depending upon water temperatures.

Starry flounders eat worms, crustaceans (especially shrimp and crabs), clams, brittle stars, and small fish, to name a few of their commonest food items. Among the fishes found in their stomachs have been sanddabs, osmerids, small herring, and small perch. Sea lions will eat netted starry founders but we doubt if they catch many on their own.

A single otolith of *P. stellatus* has been found in a Pleistocene deposit north of Arcata, California, and starry flounder scales have been found in an Indian midden near Morro Bay.

Fishery information.—Starry flounders are the most common flatfish caught by sport fishermen in central and northern California, an estimated 14,000 per year being taken during the period 1958 to 1961. About half of these are caught by pier fishermen and most of the rest by shore and skiff fishermen. They are caught throughout the entire year.

The commercial catch is made almost entirely with trawl nets, primarily from the ports of San Francisco and Eureka. Landings reached nearly 1 million pounds in 1937, but most of the time they have fluctuated between about 300,000 and 600,000 pounds. The bulk of the catch is filleted and sold fresh either under its own name or the inclusive name "sole."

Other family members.—Eighteen other righteye flounders are known to inhabit California waters, and numerous characters must be examined to identify one, although once a person is familiar with the various species, most can be recognized at a glance by noting body shape, mouth size, body color, and one or two other features. English sole can be identified by their odor, arrowtooth halibut by their dart-tipped teeth, starry flounders by their stellate scales, sand sole by some elongate filamentous dorsal fin rays, and others by similar singular or outstanding features.

Meaning of name.—*Platichthys* (flat fish) *stellatus* (starry).

Dover Sole
Microstomus pacificus (Lockington, 1879)

Distinguishing characters.—The Dover sole is easily distinguished by the abundance of slime it secretes which makes it very slippery to handle. Other helpful

Fig. 39.   *Microstomus pacificus*

[107]

characters include the straight lateral line without an accessory branch, the small gill opening which never reaches above the level of the pectoral fin base, and the large eyes and small mouth. A long loop of the intestine extends back into a "pocket" that parallels the ventral profile.

Natural history notes.—*Microstomus pacificus* ranges from the Bering Sea to the vicinity of San Quintin Bay, Baja California. It prefers areas of mud bottoms, and has been captured at depths ranging from about 100 feet to those exceeding 3,000 feet. The maximum reported length is 2 feet 4 inches. This fish, a female, weighed 10 1/3 pounds and was at least twenty years old.

Dover sole mature when they are five years old, and 12 to 13 inches long, the females being slightly longer than males. Spawning takes place in deep water from November to March, and after spawning, the adults migrate into shallower waters. The eggs are spherical and average about 1/10 inch in diameter. A 15-inch female will liberate about 37,000 to 50,000 eggs per spawning, whereas a 25-inch female contained an estimated 250,-000. Larval Dover sole are transparent leaf-shaped creatures, and they live in the pelagic environment many miles from shore.

Dover sole feed heavily upon worms and other soft-bodied invertebrates that live in and on the muddy bottoms they inhabit. Larval Dover sole are fed upon by many pelagic fishes including albacore, but no specific predator is known for adult fish.

Otoliths of *M. pacificus* have been found in several Pliocene and Pleistocene deposits in southern California, but they are not common.

Fishery information.—There is no sport fishery for Dover sole. The commercial catch is made exclusively with trawling gear, and landings are heaviest at our most northerly ports. Annual landings have ranged from about 7 to 12 million pounds during the last quarter-century. Heaviest landings are made during late spring and summer when the fish are on the inshore grounds. When they move onto the spawning grounds in winter the fishery is inhibited by water depth, stormy weather, and the distances that vessels must travel to reach the fishing grounds. The entire Dover sole catch is marketed in the form of fresh or frozen fillets.

Other family members.—For information on the nineteen kinds of righteye flounders found off California, please refer to our checklist of game and food fishes (p. 165), and to our list of references (p. 172).

Meaning of name.—*Microstomus* (small mouth) *pacificus* (Pacific, the ocean it inhabits).

Polynemidae (Threadfin Family)
Blue Bobo
Polydactylus approximans (Lay and Bennett, 1839)

Distinguishing characters.—The anchovylike nose (overhanging the mouth), two widely separated dorsal fins, and six free lowermost rays of the pectoral fin will distinguish the blue bobo from all other marine species known to California.

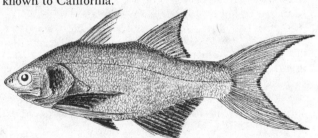

Fig. 40   *Polydactylus approximans*

Natural history notes.—The recorded range of *Polydactylus approximans* is from Monterey Bay (once) to Callao, Peru, but it is not common north of Cape San Lucas. The tiny young are pelagic, sometimes occurring several hundred miles offshore, but the juveniles and adults inhabit shallow waters near shore (and in bays and sloughs) where the bottom is sandy to muddy. The

[109]

largest we have seen was 14 inches long and weighed just under 1 pound. It appeared to be eight years old if the growth rings on its otoliths were correctly interpreted.

We have observed ripe females from June until September, but have no data on the exact spawning season, age at first spawning, number of eggs per female, or other facets of reproduction.

The few stomachs we have examined have contained mostly worms, small shrimplike crustaceans, sand crabs, and clam parts; although remains of a small anchovy were found in one stomach. The free rays beneath the pectoral fins are very sensitive to stimuli, probably physical but possibly chemical also, and they are used rather like radar in locating food. The inch-long pelagic stages are heavily fed upon by tunas, jacks, and other predators that are found at the surface in the offshore environment. The shallow-water, bottom-dwelling juveniles and adults undoubtedly are preyed upon by an assortment of fish-eating birds, mammals, and finny relatives, judged by the rapidity with which they will accept a fresh bobo when it is offered as bait or food.

Fishery information.—Blue bobos rarely stray into our area except during warm-water years. The thirteen months between August 1940 and August 1941 appear to have produced the best "annual" catch known. During this period, eight, 10- to 12-inch-long fish were caught off California, seven between San Pedro and San Clemente, and one in Monterey Bay. They are quite tasty when fried.

American tuna fishermen often used bobos for bait when chumming and catching tropical tunas with hook and line, but since the tuna clippers have been converted into purse seiners, bait fishes are no longer used. These tuna fishermen are responsible for the name "bobo," which is a distortion of the Spanish vernacular.

Other family members.—One other member of the family has been taken off southern California, but it is

[110]

even rarer than the blue bobo. The two species are easy to distinguish by checking color as well as the number of free rays beneath the pectoral fin.

1. If there are 6 free rays, and the fish is silvery-blue it is a blue bobo.
2. If there are 9 free rays, and the fish is silvery-yellow it is a yellow bobo.

Meaning of name.—*Polydactylus* (many finger) *approximans* (approaching, in reference to its similarity to a previously named species).

Pomacentridae (Damselfish Family)
Blacksmith
Chromis punctipinnis (Cooper, 1863)

Distinguishing characters.—The blacksmith is the only perch-shaped fish in our waters with a dusky-blue body that is profusely covered with black spots on the posterior half. Black spots also cover much of the second dorsal fin and the tail. Juveniles, up to about 2 inches in length, are purplish-blue anteriorly (to about a level with the first anal fin spine), and yellowish from there back. The upper and lower margins of the tail fin are iridescent blue on these little juveniles.

Fig. 41.   *Chromis punctipinnis*

Natural history notes.—*Chromis punctipinnis* ranges from Monterey Bay, California, to Turtle Bay, Baja California. It is an inhabitant of shallow rocky areas, including around breakwaters, and is especially abundant in kelp beds around the Channel Islands. It

[111]

seldom is seen singly; an apt term for an aggregation of blacksmiths would be a "swarm."

They are reported to attain a length of 12 inches, but the largest we can find information on was an 11 1/2-inch female taken at Carmel Bay in January 1962. Neither this fish, nor another from Monterey Bay that was almost identical in size, weighed quite 1 pound. The Carmel Bay blacksmith was seven years old, judged by growth zones on its otoliths, and it had first spawned when it was three. Some blacksmiths mature when they are about 5 1/2 inches long and two years old.

During the spawning season, which usually runs from June through August, a male blacksmith selects a nesting site under a rock ledge or in small caves under rocks (in 20 to 80 feet of water), and then cleans off all encrusting growths and algae. Once this has been accomplished he finds a gravid female, herds her into his "nesting cave," and fertilizes the eggs that she spawns. After the female departs, the male remains at the entrance to the nest site and guards the eggs until they hatch. It is believed that some nests contain eggs from three or four females that the male has herded into his cave. The eggs are attached by adhesive filaments to each other and to the cleaned area in the cave roof. At spawning, they are salmon pink in color and oblong in shape, being slightly longer than one-twentieth of an inch each. A single nest was estimated to contain over 600,000 eggs. Information is lacking on the time it takes for the eggs to hatch. Some larvae and juveniles are encountered many miles offshore accompanying drifting masses of kelp. We seriously doubt that all blacksmith undergo a pelagic stage, however.

Their principal food appears to be small shrimplike crustaceans that inhabit the kelp-bed environment. They also are cannibalistic insofar as their own eggs and newly hatched larvae are concerned, and they will eat larvae of other fish as well as larval squid. Juvenile and adult blacksmiths are preyed upon by an assortment of large fishes that frequent shallow kelp beds and rocky areas, and they have been found in sea lion and cormorant stomachs.

Fishery information.—Few sport anglers fish specifically for blacksmiths, but fishermen who use small hooks while fishing rocky areas of southern California will catch fair numbers. Novice skindivers often spear blacksmiths because swarms of them are everywhere in the kelp beds which are favorite skindiving spots. They are not a desirable fish because of their small size, and probably fewer than 500 are carried home each year by sport fishermen.

Small quantities are taken by commercial fishermen incidental to other fishing activities. If these are saved, they are marketed fresh and reported in catch statistics as "miscellaneous perch." Probably less than a ton a year is utilized by the markets.

Other family members.—The garibaldi is the only other damselfish known to our waters. This large-scaled, bright orange fish is impossible to mistake. Juvenile garibaldis (to about 4 inches) are blotched with iridescent blue.

Meaning of name.—*Chromis* (the ancient name for some fish) *punctipinnis* (spot fin).

Pomadasyidae (Grunt Family)
Sargo
Anisotremus davidsonii (Steindachner, 1875)

Distinguishing characters.—Because of its size, color, and shape the sargo might be mistaken for a perch, but it differs from all the embiotocids in several external features. The anal fin of the sargo contains nine to eleven soft rays (no surfperch has fewer than thirteen), and the second anal spine of the sargo is strong (heavy) and as long as any of the soft anal rays (anal fin spines of embiotocids are relatively weak and only about half as long as the rays). In addition, sargos will make piglike grunting sounds when hauled from the water.

Natural history notes.—*Anisotremus davidsonii* ranges from Monterey Bay (based upon a 1960 capture)

[113]

to Magdalena Bay, Baja California, and occurs again in
the northern Gulf of California. The sargo was introduced into Salton Sea (from the Gulf) in March 1951
and has become one of the most abundant sport species
there. In coastal waters of southern California they are
most abundant in and around kelp beds and other shallow environment where there is a rocky or combination
rock and sand bottom. They are also commonly seen
around pier pilings and similar "solid relief" in shallow
bays as well as on the outer coast. They are usually in
small loose schools or aggregations and have been observed into depths of 130 feet.

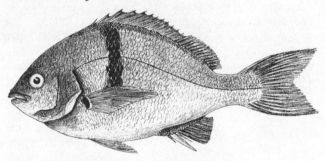

Fig. 42.   *Anisotremus davidsonii*

Sargos are known to reach lengths in excess of 20
inches (probably to 23 inches) and weights of 5 to 6
pounds, but none of these "giants" has had its vital statistics taken. The largest we have seen was a 17 2/5-
inch female that weighed 3 7/10 pounds. A slightly
smaller fish was fifteen years old, judged from growth
rings on its scales and otoliths, so at maximum size they
might be older than twenty.

Spawning occurs in late spring and early summer, but
we have no information on age at first spawning, number of eggs spawned per season, and other details of reproduction. The ovaries of a ripe female will often
make up 20 to 25 percent of the fish's total weight,
however.

[114]

Sargo stomachs have contained mostly crustaceans (amphipods, isopods, shrimp, crabs, and barnacles), mollusks (kelp scallops, clam siphons, and small snails), and polychaete worms. We do not know of any specific predators on sargos, but sea lions and some porpoises probably catch and eat medium-sized individuals.

A fossil otolith of *A. davidsonii* was found in a Pleistocene deposit at Playa del Rey that was laid down over 20,000 years ago.

Fishery information.—Sargos are caught by anglers fishing from rocky shores, breakwaters, piers, and docks, both inside bays and along the outer coast. They are taken by skiff and small boat fishermen around Santa Catalina Island and are a prime target for spearfishermen. A recent survey of marine sportfishing in southern California indicated an estimated 15,000 sargos were being taken annually.

The commercial catch is made primarily with gill nets as a by-product to other fisheries, although a large winch-operated lift net has been used successfully during some years. Sargos are marketed as "perch" in the fresh fish trade but have never been important. There probably have been few years when even a thousand pounds were sold.

Other family members.—One other member of the family (the salema) is found in our waters during some years, but it is easily distinguished from the sargo. The salema differs from the sargo in body shape, eye diameter, dorsal fin configuration, and color, but only the last two characters will be used in the following simple key.

1. If the two dorsal fins are separate, and there are 6 to 8 brassy horizontal stripes on the sides it is a salema.
2. If the spiny and soft dorsal fins are broadly joined, and there is a dusky vertical bar on the back and side it is a sargo.

Meaning of name.—*Anisotremus* (unequal aperture, pertaining to the pores at the chin) *davidsonii* (for Professor George Davidson, an early-day astronomer at the California Academy of Sciences).

Salmonidae (Salmon Family)
King Salmon
Oncorhynchus tshawytscha (Walbaum, 1792)

Distinguishing characters.—Young salmon sometimes are mistaken for trout, but trout have a squared-
off tail fin and twelve or fewer rays in the anal fin. Juvenile king salmon have large, oval-shaped parr marks on
their sides and the spaces between the parr marks are
narrower than the horizontal distance across one parr
mark. Adult king salmon have small black spots on the
upper body, dorsal fin, and on the entire tail fin. In addition, their gums (area around the teeth in both jaws) are
black. These characters are sufficient to distinguish
them from all other salmons found in the ocean off our
coast.

Fig. 43.   *Oncorhynchus tshawytscha*

Natural history notes.—*Oncorhynchus tshawytscha*
ranges throughout the north Pacific from Japan to the
Bering Sea and south from there to San Diego. It is a
pelagic schooling fish, but a "school" of salmon seldom
contains more than 100 or so individuals in a loose aggregation. A 126 1/2-pound king salmon appears to be
the all-time record, but only rarely is a 100-pounder encountered, and 60-pounders are considered unusual. The
average ocean-caught fish will weigh about 10 pounds,
while the average-sized spawner will be about 20. A 92-
pound king salmon was 4 feet 10 inches long.

Some male king salmon become sexually mature
when they are a year old, and about 6 inches long.

[116]

Nearly all kings spawn when either three or four years old, but some will not spawn until five or six. Seven years appears to be maximum for king salmon, but well over 70 percent never reach five. All but the precocious year-old males die after spawning.

King salmon undertake both spring and fall spawning runs, with heaviest runs in our state entering the Sacramento River system (from 100,000 to 500,000 spawners have used the Sacramento River per year since about 1950). A Sacramento River king salmon female will produce an average of 6,000 eggs. These will hatch in fifty to sixty days depending upon water temperatures at the time (the warmer the temperature the earlier the hatching). In California, most king salmon young will migrate down the river, tail first and at night, until they reach the ocean. Losses due to predation, irrigation water diversion, and pollution are very heavy during their migration to the ocean, and mortality does not stop even then. They grow very rapidly in the sea (compared with their stream growth) and often travel many hundreds of miles before returning to their home stream as spawning adults from two to seven years later.

While in the ocean they feed heavily upon an assortment of fishes, crustaceans, and squids. An examination of more than 1,000 stomachs of ocean-caught fish (near San Francisco) revealed that six species made up more than 92 percent of the diet. These six "species" were anchovies, rockcods, euphausiids, herring, squid, and larval crabs. King salmon will be taken from nets by seals and sea lions, but it is doubtful if they prey upon them under natural conditions.

An otolith of *O. tshawytscha* was found in a Pleistocene exposure north of Arcata that probably was deposited more than 20,000 years ago. Coastal Indians fed heavily upon king salmon, most of which were probably caught in rivers and streams rather than at sea.

Fishery information.—From Monterey north king salmon are probably the most highly prized species of all those sought by the ocean angler. Most are caught by trolling an aritificial lure 20 to 60 feet beneath the surface, but many are taken between the surface and 250 feet or deeper while drift fishing with live bait. The sport catch, made primarily from partyboats and skiffs, has ranged from 5,000 to 129,000 fish during the past quarter century. The best catch was made in 1955.

Ocean-caught king salmon are superior in quality, flavor, and condition to most river-caught fish, and their flesh is a deeper color. The commercial catch usually has averaged around 10 million pounds per year since the Second World War, but it has ranged from about 5 1/2 to more than 13 million. King salmon are utilized almost exclusively in the fresh state and are excellent barbecued, broiled, smoked, or baked.

Other family members.—Five other members of the family are found in the ocean off California: four salmons and the steelhead trout which is nothing more than a sea-run rainbow trout. Adult salmonids are not easy to tell apart, but the following characters should work on freshly caught fish.

1. If the tail fin is squared off and there are 12 or fewer anal fin rays it is a steelhead trout.
2. If the tail fin is indented (slightly forked) and there are more than 12 anal fin rays it is a salmon.
   a. If the salmon has distinct black spots on the back and tail and
      i. the gums (around teeth) are black it is a king salmon.
      ii. the spots on the back and tail are small, and those on the tail are on the upper half only it is a silver salmon.
      iii. the spots on the back and tail are large (eye sized), and those on the tail cover the entire fin it is a pink salmon.
   b. If there are no distinct spots on the back (sometimes lightly sprinkled with pinpoint speckles) or tail, and
      i. the mouth lining is white it is a chum salmon.
      ii. the mouth lining is black it is a sockeye salmon.

Meaning of name.—Oncorhynchus (hook snout) *tshawytscha* (a distortion of the vernacular name in Alaska and Kamchatka).

Sciaenidae (Drum Family)
White Seabass
Cynoscion nobilis (Ayres, 1860)

*Distinguishing characters.—*The white seabass is the only fish in our waters with a raised ridge along the midline of the belly. Juveniles, up to about 18 inches, have four or five fairly broad dark vertical bars on their sides, but adults are bluish above and silvery white on the lower sides and bellies.

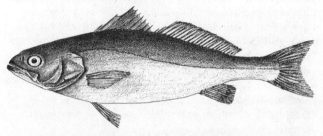

Fig. 44.   *Cynoscion nobilis*

*Natural history notes.—*During some years *Cynoscion nobilis* has been caught as far north as Juneau, Alaska, but its usual range extends from about San Francisco to Magdalena Bay and it is found also throughout the northern Gulf of California. An 83-pound 12-ounce fish has been the recognized world record for many years. The age of this individual is unknown, but 40-pounders often are twenty years old or older.

A fair percentage of the females are mature and will spawn when they are three years old, and almost all will spawn at age four. We have no information on the number of eggs each female spawns during a season, nor on length of time till hatching. Spawning occurs during late spring and throughout the summer months,

[119]

and very tiny young can be netted in relatively shallow nearshore embayments during much of this period. Large adults are equally at home in the surf zone or in 350 to 400 feet of water, but they seem to prefer depths of 75 to 150 feet.

Juvenile white seabass are probably eaten by whatever large predators can catch them, and they appear to be especially affected by industrial and domestic pollutants. Adults have few enemies except man, although sea lions and sharks will wreak havoc with netted fish. Fish — especially anchovies, sardines, and small mackerel — and squid make up the bulk of food eaten by white seabass.

Fossil otoliths of *Cynoscion nobilis* have been found in several southern California Pleistocene deposits, and in a Pliocene exposure at San Diego. Some of these probably are 10 to 12 million years old. White seabass otoliths have also been found in many coastal Indian middens, and a necklace containing more than twenty-five otoliths interspaced with shells of olive snails was dug up at an Indian campsite on San Nicolas Island.

Fishery information.—White seabass are highly prized by sport fishermen, and as many as 65,000 have been caught during a single year (1949) by partyboat fishermen alone. Skiff fishermen, skindivers, and other anglers also take a heavy toll of white seabass. Rod and reel fishermen do best with live anchovies for bait, but medium-sized sardines, small mackerel, and squid often make equally good bait. Artificial lures made of bone, metal, or plastic in a variety of shapes and colors are also very popular. Sometimes fishing is better at night than during daytime.

White seabass are best in steak form, and most of the commercial catch is sold as seabass fillets — they are never canned. Round haul or encircling nets (purse seines, lamparas, etc.) were made illegal in 1940 for catching white seabass, so the bulk of the catch since then has been made with gill nets. About a million pounds have been landed annually during the past fifty years but as few as 245,000 pounds were taken by commercial fishermen in 1944 and nearly 3.8 million pounds were sold in 1959.

Other family members.—Seven other members of the drum family are found in our coastal waters and two additional kinds are abundant in Salton Sea. A few simple characters will distinguish the eight sciaenids that live off our coast:

1. If the mouth is terminal and
   a. there is a raised ridge on the midline of the belly it is a white seabass;
   b. the two dorsal fins are widely separated it is a queenfish;
   c. if neither of these characters fits it is a shortfin corvina.
2. If the mouth is subterminal, a chinwhisker is present, and
   a. there are rust-colored wavy streaks on the sides and back it is a yellowfin croaker;
   b. if the sides and back are dark grayish blue it is a California corbina.
3. If mouth is subterminal, chinwhisker is lacking, and
   a. there are 9 to 11 dorsal spines, and
      i. 21 to 24 soft dorsal rays it is a spotfin croaker.
      ii. 25 to 28 soft dorsal rays it is a black croaker.
   b. there are 12 to 15 spines in the first dorsal fin it is a white croaker.

Meaning of name.—*Cynoscion* (dog *Sciaena*, in allusion to its doglike canines and presumed resemblance to one of the Mediterranean drums) *nobilis* (noble).

California Corbina
Menticirrhus undulatus (Girard, 1854)
Distinguishing characters.—The California corbina is easily identified by its characteristic shape in combination with a subterminal mouth, a single fleshy barbel or chinwhisker on the lower jaw, and the dusky grayish-blue color on sides and back.

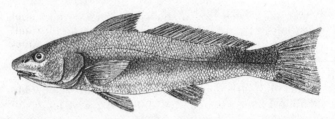

Fig. 45.   *Menticirrhus undulatus*

Natural history notes.—Menticirrhus undulatus ranges from Point Conception to just north of Magdalena Bay, Baja California. Very large fish usually occur singly, but their habitat preferences and habits are otherwise similar to those of smaller individuals. These fish usually travel in small groups (perhaps six to twenty-five individuals) and are found in water ranging from a few inches deep in the swash zone of the surf to depths of about 45 feet. They live almost exclusively over a sandy or firm sandy-mud substrate, usually along the open coast but also in shallow embayments and sloughs where conditions are satisfactory. Apparently they move offshore during winter months and for spawning, but the greatest coastal movement reported during a tagging study was fifty-one miles.

Although an 8 1/2-pound California corbina was reported by a fisherman, the largest verified fish weighed 7 pounds 4 ounces and was 28 inches long. We have no information on the age of this large fish, but the oldest of 1,724 specimens examined in an age and growth study was eight years old; this fish, a female, was just under 20 inches long.

Most males are mature when they are two years old and about 10 inches long, whereas females are a year older and 3 inches longer before most of them are ready to spawn. Spawning takes place from June to September but is heaviest during July and August.

Sand crabs are definitely the preferred food of the California corbina, but when beds of bean clams form a

[122]

pavement in the intertidal zone, great quantities of these are eaten by the corbinas in the area. Other small crustaceans (amphipods and mysids) as well as polychaete worms and grunion eggs have also been found in their stomachs. Corbina remains have been found in the stomachs of several predatory species including a Pacific bottlenose dolphin, a cormorant, and an angel shark.

Fossil otoliths of *M. undulatus* have been found in a Pleistocene deposit at Playa del Rey (near Los Angeles International Airport); these have been estimated to be more than 20,000 years old.

Fishery information.—It has been illegal to take California corbina with nets since 1909, or to buy or sell them since 1915. Surf and pier fishermen in southern California are especially pleased to catch corbina, and they will spend days bragging about any that exceed 3 or 4 pounds in weight. *M. undulatus* is reported to make up 17 percent of the surf fisherman's bag in southern California, and a study conducted in the early 1960s showed that about 38,000 had been caught that year — 30,000 by shoreline fishermen and most of the rest by pier fishermen.

Other family members.—Refer to p. 121 for this information.

Meaning of name.—*Menticirrhus* (chin barbel) *undulatus* (waved).

Scomberesocidae (Saury Family)
Pacific Saury
Cololabis saira (Brevoort, 1856)

Distinguishing characters.—The sharply pointed (almost beaked) snout, slender body that is roundly triangular in cross section, fins placed well back on the body, and presence of dorsal and anal finlets are more than sufficient to identify a Pacific saury. The dark blue to green back and brilliant silver flanks and belly are also very characteristic.

[123]

Fig. 46. *Cololabis saira*

Natural history notes.—Cololabis saira is a schooling fish of the open north Pacific Ocean, ranging on our coast from Alaska to about Cape San Lucas, Baja California. Juveniles and larvae predominate in waters south of about Monterey Bay, while large adults are present in commercial quantities north of there and offshore. The species is reported to attain a length of 14 inches, but most fish observed off our coast are smaller than 10 inches. The largest we have seen was a 13-inch female spit up by an albacore in December 1957. This fish weighed just over 4 ounces, and was three years old according to growth rings on its scales and otoliths. Its stomach was empty, but other stomachs we have examined contained small crustaceans almost exclusively.

Japanese scientists report that some female Pacific sauries mature when two years old and that all are mature when three. A single female will liberate an average of about 1,800 eggs each time it spawns, and one fish will spawn six or seven times per season, commencing in December and tapering off in the autumn. Maximum age appears to be five years, but few individuals survive past age four. The largest of 20,000 fish examined by the Japanese scientists was 15 inches long.

Large sauries are often heavily infested with external parasites. Hairy plumelike copepods, *Pennella*, may be embedded somewhere along the sides, and there are frequently several roundish open sores or scars which have been caused by another parasitic copepod, *Caligus*. Both these detract from their sale as a fresh product for human consumption.

Sauries are fed upon by a wide variety of fish-eating predators. Squids, especially *Dosidicus*, often have their stomachs crammed with the remains of juvenile sauries

[124]

(identifiable from otoliths), and at times albacore and marlins feed upon them almost exclusively. Sharks, marine mammals, and sea birds also take a heavy toll.

Fishery information.—There is no sport fishery for the Pacific saury, and while commercial landings are heavy at times, they are sporadic and unpredictable. A small amount was canned experimentally in the Monterey area in 1947, and in 1957 considerable tonnage was delivered to a Wilmington, California, canner for processing as pet food, but landings during other years have not been worthy of note. There would be a much greater interest in fishing for sauries commercially if it could be done profitably. Unfortunately there is little demand for fresh or canned sauries in this country, but even if the product could be readily sold the fishermen would be faced with purchasing and constructing fine-meshed nets, traveling great distances for a load of fish, and perfecting capture techniques for these elusive, fast-moving creatures. The Russians are very much interested in the saury populations off central and northern California, and probably will be fishing for them within a few years. The Pacific saury is highly prized by the Japanese and landings off their coast have exceeded 300,000 tons during some years; the fish are caught primarily with dip nets at night after they have been attracted to a strong electric light.

Other family members.—No other member of the family is known within a thousand miles of our shores.

Meaning of name.—*Cololabis* (defective forceps, with reference to the jaw shape) *saira* (one of two Japanese vernaculars for this species).

Scombridae (Mackerel Family)
Albacore
Thunnus alalunga (Bonnaterre, 1788)

Distinguishing characters.—Albacore and bigeye tunas are the only two tunas with very long pectoral fins, which reach past the point of insertion of the anal fin. They can be separated immediately if the two are side by side, but most characters to distinguish them are relative, so if only one of the two is at hand there may be some difficulty in identification. The albacore has a slightly shorter head (back edge of gill cover fails to reach a perpendicular beneath the first dorsal insertion), the vent is round (oval in the bigeye), and the underside of the liver is solidly streaked with blood vessels (faint marginal striations in the bigeye).

Fig. 47. *Thunnus alalunga*

Natural history notes.—Schools of *Thunnus alalunga* occur seasonally along the coast of California. They have a distinct preference for temperate waters, but they have been taken all the way from Clarion Island (tropical Mexican waters) to well up the Alaskan coast. Mostly though they range from about Cedros Island, Baja California, to off the coast of Oregon, between about mid-June and the end of October. During the rest of the year they are traveling between California and Japan with round-trip tickets. Other stocks of albacore are found south of the equator in the Pacific Ocean, as well as throughout the Atlantic. They prefer deep blue oceanic water, rarely being seen in greenish or dirty inshore areas.

The largest sport-caught fish on our coast appears to be a 66 1/2-pounder, although a 76-pounder has been taken off California by a commercial fisherman. The largest we could find a record for is a 93-pounder caught on longline gear in the mid-Pacific. Albacore

[126]

ages have been determined by studying scales, length frequencies, and tag recoveries. From these data it was determined that albacore add from 6 to 8 pounds per year for the first six or seven years and somewhat less thereafter. Thus, a six-year-old fish (about 40 inches long to the fork of the tail) would weigh around 45 pounds. A 70-pound albacore probably would be ten to twelve years old.

The nearest albacore spawning grounds are probably in the mid-Pacific, and spawning appears to take place between January and June, prior to the time the fish migrate into our coastal waters. Available data indicate that female albacore first spawn when they are 34 to 36 inches long (to the fork of the tail). The ovaries of a fish this size are estimated to contain over a million eggs, each about a twenty-fifth of an inch in diameter when spawned. Hatching apparently takes place within a matter of days.

The food of the albacore varies depending upon whether they have been feeding at the surface or deep and what items are easiest to obtain at the time, place, and depth the albacore is in a feeding mood. Small fishes comprise the bulk of their diet, but cephalopods (primarily squids) and small shrimplike organisms are also very important.

Albacore were being caught for food by some of the Indians inhabiting the southern California coast when the first Spanish explorers showed up, but there are no fossil records of *Thunnus alalunga*. This is probably a reflection of their pelagic habits rather than the fact that they did not exist.

Fishery information.—Sport fishermen today consider albacore as one of the most desirable of all game fishes in our waters for both recreation and food, but they have not always enjoyed this popularity. In the late 1800s and early 1900s they were considered a nuisance by many sport fishermen, especially those seeking the large, tackle-breaking bluefin tuna. By 1930, though, this attitude had changed. The best albacore sport fishing season on record was 1962 when over 229,000 fish were landed by partyboat fishermen, the poorest season (1959) yielded forty albacore to the sport fishery.

The commercial albacore fishery dates back to about 1900, but catches were sporadic and often poor until 1935. Since 1935 catches have been generally heavy as the albacore fleet has followed the fish in their northward migration along our coast. Much of the success that albacore fishermen enjoy today is the direct result of research efforts of biologists employed by the California Department of Fish and Game. Tags developed by these biologists demonstrated a trans-Pacific migration and helped to pinpoint albacore approach routes. Catch data were analyzed, and an ocean temperature relationship was found that has proven helpful in locating areas of greatest fish concentration. Continuing studies have revealed numerous other facets of albacore behavior and biology, and all are proving useful in locating and catching these elusive migrants. The best commercial season on record was 1950 when 61 million pounds were taken by California fishermen; in the poorest season (1933) only 500 pounds were landed.

Other family members.—Although fifteen scombrids have been reported from our coast, only four of these belong to the genus *Thunnus*. Three of these four (albacore, bluefin, and bigeye tunas) prefer temperate waters, but the yellowfin tuna is a tropical species. None lives off our coast on a year-round basis.

Meaning of name.—*Thunnus* (ancient name for the tunny — our bluefin tuna) *alalunga* (long wing, in reference to the elongate pectoral fins).

Pacific Mackerel
Scomber japonicus Houttuyn, 1782
Distinguishing characters.—Pacific mackerel are distinguished from all other fishes in our waters by the mackerellike or tunalike body, widely separated first

and second dorsal fins, five dorsal finlets, and five anal finlets.

Natural history notes.—Scomber japonicus is a schooling fish that in the eastern north Pacific has been found from the Gulf of Alaska to Banderas Bay, Mexico. Other populations are found off the west coast of South America, and in the western Pacific in waters adjacent to Japan.

Fig. 48.   *Scomber japonicus*

The largest Pacific mackerel on record, nearly 25 inches long and weighing 6 1/3 pounds, was probably a freakish giant, but since no information is available on its age, this is only speculation. The oldest of more than 30,000 individuals for which ages have been determined was twelve years old. A fish that age would be about 18 inches long and weigh just over 2 pounds. Many fish will spawn when they are two years old and all will spawn when they are three.

Spawning is usually at a peak during March through May, although some spawning may occur a month or two earlier and later depending upon water temperatures and other environmental conditions. Pacific mackerel eggs are about one twenty-second of an inch in diameter (1.15 millimeters) and drift in the upper water layers until they hatch in four or five days.

Pacific mackerel feed mostly upon small fish (anchovies and such) and euphausiids, but they will eat whatever other bite-sized organisms they encounter in their environment, especially juvenile squid. Porpoises, sea lions, yellowtail, giant sea bass, white seabass, marlin, sharks, and many other large predators will feed on Pa-

[129]

cific mackerel whenever the opportunity arises. Diseases and parasites do not seem to be a hazard, although parasitic roundworms frequently are found in the intestinal tract of a mackerel.

Remains of S. *japonicus* have been found in several coastal Indian middens, some having been discarded by these primitive fishermen 5,000 years ago. Otoliths are present, though not abundant, in several Pleistocene and Pliocene deposits of southern California, and skeletal imprints and otoliths of *Scomber* (possibly S. *japonicus*) have been found in 25 million-year-old Miocene outcrops.

Fishery information.—Some sport fishermen will actually seek out Pacific mackerel for their fighting qualities on very light tackle, for the table, or for use as bait, but most fishermen regard them as trash and treat them accordingly. Catch figures are available only for the partyboat fishery and these show a high of 315,000 Pacific mackerel during a single year. Since many are taken from barges, skiffs, private cruisers and yachts, piers, docks, and breakwaters, the partyboat catch during any one year probably represents only half or less of the total sport catch.

Mackerel canning techniques were not perfected until 1928, so the demand (and catch) until then had not been great. The catch jumped from about 5 million pounds in 1928, to 35 and 58 million during the next two years and then dropped again during the depression years of 1930 through 1933. The commercial catch reached a peak of 146 million pounds in 1935 and has been declining at an irregular rate ever since. The fishery has been in sad shape during most years since the mid 1950s, and only an occasional good hatch has kept the mackerel from vanishing completely.

Other family members.—In all, fifteen scombrids have been taken off our coast. One (*Allothunnus*) has been captured only once, four have been seen fewer than five times (*Auxis thazard, Euthynnus yaito,*

Scomberomorus sierra, and *Lepidocybium)* and two, the Monterey Spanish mackerel and bigeye tuna *(Scomberomorus concolor* and *Thunnus obesus)* are still considered rare in our waters. Seven of the other eight species occur seasonally off California, with only Pacific mackerel a year-round resident (bonito share the honor during some years but not during all). To distinguish the various species one must consider fin positions and lengths, count gill rakers and fin rays, examine livers, measure or count scale rows, check backs and bellies for stripes, and examine other anatomical features.

Meaning of name.—Scomber (ancient name of the mackerel) *japonicus* (Japan, where the species was first described).

Scorpaenidae (Rockcod Family)
Cowcod
Sebastes levis (Eigenmann and Eigenmann, 1889)

*Distinguishing characters.—*The light pink body crossed by several broad dusky bars, in combination with the wide space between the lower margin of the eye and the upper lip, will distinguish the cowcod from the other fifty-one members of the genus found off California.

Fig. 49.   *Sebastes levis* (juvenile)

Natural history notes.—Sebastes levis ranges from Fort Bragg, California, to Ranger Bank, Baja California.

[131]

Juveniles 3 to 5 inches long are often found on sandy bottom areas in 60 to 100 feet of water, but large adults seem to prefer rocky areas at depths of 500 to 800 feet. Many 30- to 38-pounders have been reported in newspapers and magazines, but the largest we have weighed on approved scales was 28½ pounds. A few hours prior to our weighing, this same fish had won a prize as a 36-pounder after having been weighed at dockside. (The dockside scales were overweighing all fish by more than 25 percent!) At 37 inches, this 28½-pounder was the longest cowcod on record.

We have no information on age or size at maturity, or on maximum age. A fish 13 inches long weighed just under 1¼ pounds and was three years old. It was not mature, but one 24 inches long weighing 7½ pounds (7 years old) was nearly ready to spawn. A 20-pound cowcod was estimated to be more than eighteen years old.

As with all members of the genus *Sebastes*, fertilization is internal, and the embryos develop within the ovaries of the female until they are ready to hatch. Cowcod spawn during winter and early spring months, and a large adult is believed to release more than 2 million tiny embryos at one spawning.

Juvenile cowcod stomachs have contained mostly small crustaceans (shrimp and crabs), but larger fish require larger food items, and their stomachs have contained mostly fish, octopi, and squid. Adults probably have few enemies other than man, although on rare occasions a large individual will be caught that has healed scars on its body from shark bites.

Otoliths of *Sebastes* are abundant in Pliocene and Pleistocene deposits along the California coast, but those of *S. levis* have not yet been identified.

Fishery information.—Sport fishermen actively fish for cowcod because of their large size and the opportunity to win either the boat jackpot or some prize offered by the landing operators, a sport fishing magazine, or some industry. There are no figures on the statewide annual catch of cowcod, but it undoubtedly amounts to several thousand if reliability can be placed in the daily newspaper accounts based on information furnished by various partyboat landings.

Commercial setline fishermen catch cowcod incidental to fishing activities conducted for other market species, particularly vermilion rockcod. A two-man boat will catch about 15 "cows" for each 1,000 "reds." The "cows" will average about 15 pounds each. Commercial rockcod catches are not broken down by species, so it is impossible to estimate the annual take of cowcod.

Other family members.—Fifty-six kinds of scorpaenids belonging to four genera inhabit the marine waters of our state. There are fifty-two species of *Sebastes,* two of *Sebastolobus,* one of *Scorpaena,* and one of *Scorpaenodes.* For one who is unfamiliar with these fishes, it is necessary to count scales, gill rakers, fin rays, and head spines, and to measure numerous body parts and appendages in order to identify one. Body color is perhaps the most important single character, however.

The four genera can be distinguished by the following:

1. If there are 12 dorsal fin spines it is a California scorpionfish, *Scorpaena guttata.*
2. If there are 13 dorsal fin spines and palatine teeth are present it is one of the fifty-two kinds of rockcod, *Sebastes.*
3. If there are 13 dorsal fin spines and there are no teeth on the palatine it is a rainbow scorpionfish, *Scorpaenodes xyris.*
4. If there are 15 or 16 dorsal fin spines it is one of two kinds of thornyheads, *Sebastolobus.*

Meaning of name.—*Sebastes* (magnificent, probably in reference to the brilliant colors) *levis* (capricious or fantastic).

Bocaccio
Sebastes paucispinis Ayres, 1854
Distinguishing characters.—The bocaccio can be

easily distinguished by its color (olive-brown above and pinkish-orange on the sides and belly), rather smooth head (lacking in spines), and very long upper jaw (the maxillary bone reaches past a point below the rear margin of the eye).

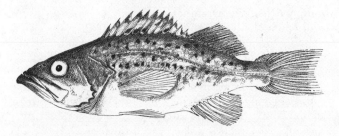

Fig. 50.   *Sebastes paucispinis* (juvenile)

Natural history notes.—Sebastes paucispinis has been captured between Kruzof Island, Alaska, and Sacramento Reef, Baja California. Juveniles often are abundant in shallow water just outside the surf zone, in both rocky and sandy bottom areas. Adults inhabit greater depths and are expecially abundant in 250 to 750 feet where the bottom is firm sandy-mud, rubble, or solid rock.

Bocaccio are reliably reported to reach a length of 34 inches, but not many fish of that size have been seen. A 32-inch female captured off Paradise Cove, California, in 1961 weighed 14 pounds, but its age was not determined. A slightly larger individual was found to be thirty years old, however.

Some bocaccio are mature when 14 inches long and three years old, and about half are mature at 16½ inches and four years, but some do not spawn until they are six. Spawning occurs from December through April and a 15-inch female will liberate about 20,000 embryos. A fish 30 inches long will spawn over 2 million embryos during a single season. A newly-spawned larva appar-

[134]

ently does not absorb the yolk from its egg stage for a period of eight to twelve days.

Before completing their first year young bocaccio start eating other fish smaller than themselves, in addition to an assortment of small crustaceans. Large bocaccio feed on smaller rockcod, sablefish, anchovies, lanternfish, midshipmen, pelagic red crabs, and squid. Bocaccio remains have been found in the stomachs of whales, porpoises, sea lions, and sharks. Man is their greatest enemy, however.

Fossil otoliths of S. *paucispinis* have been found in several Pliocene and Pleistocene deposits in California, and bocaccio apparently were used for food by coastal Indians for several thousand years before the state was settled by "civilized" man.

Fishery information.—Bocaccio sometimes are taken in great numbers by sport fishermen, but they seldom are the preferred species. A fishery survey north of Point Arguello to the Oregon border revealed that of nearly 1 million rockcod taken annually by sport fishermen, only 44,000 were bocaccio.

Rockcod have been important to the economy of California since 1875, at least. Setlines and otter trawls are the primary gear used in the commercial fishery. Rockcod landings (all species combined) have fluctuated widely since the 1940s, reaching slightly less than 15 million pounds during some years and about one-third this during others. Bocaccio may contribute from 10 to 20 percent of this total, but there is no valid information on their importance. Most are marketed as fresh fish, but some are processed for pet food or fed to animals at fur farms.

Other family members.—For information on other family members please refer to p. 133, to our checklist of game and food fishes (p. 168), and to our list of references (p. 172).

Meaning of name.—Sebastes (magnificent) *paucispinis* (few spines).

Serranidae (Sea Bass Family)
Giant Sea Bass
Stereolepis gigas Ayres, 1859

Distinguishing characters.—The giant sea bass is the only member of this family living off California that has more spines (11) than soft rays (9 or 10) in its dorsal fins. Young fish, up to about 6 inches, have perch-shaped bodies, are brick-red in color, and have six irregular rows of black spots on each side. These spots are still visible on the sides of large adults when alive and viewed in their natural surroundings, but they fade soon after death.

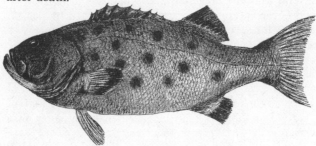

Fig. 51.   *Stereolepis gigas*

Natural history notes.—*Stereolepis gigas* has been taken between Humboldt Bay and the tip of Baja California and throughout much of the Gulf of California, but its occurrence north of Point Conception has been sporadic and unpredictable. Large fish prefer rocky-bottom habitat just outside kelp beds and along drop-off areas where the water is 115 to 150 feet deep. The large fish are most abundant around San Clemente, Santa Catalina, Santa Barbara, San Nicolas, Anacapa, and Santa Cruz islands, and upcoast from Oceanside. Small fish (up to 30 pounds at least) will be found often in sandybottom areas and in and around kelp beds in depths of 40 to 70 feet.

There is a good chance that a 600-pounder will be caught someday, but the current record giant sea bass is

[136]

a 557-pound fish caught in 1962. One this size might be ninety or one-hundred years old, but there is no reliable information to support such a theory. The oldest fish for which a valid age could be determined was seventy-two to seventy-five years old, and it weighed 435 pounds. Giant sea bass apparently do not mature until they are eleven to thirteen years old (at a weight of 50 to 60 pounds). A 320-pounder (around fifty years old) contained ovaries weighing 47 pounds in which there were an estimated 60 million eggs. Spawning takes place from June through September and during this period while the spawning adults are in relatively shallow depths they are very vulnerable to fishermen and skindivers who know where to find them. A few skindivers and sport fishermen have been known to spear or hook twenty to fifty fish each during a single spawning season, showing they have no concern for sportsmanship or the basic principles of conservation.

Inch-long young appear in December, and by February they have increased to 2 inches. At one year, giant sea bass attain a length of 7 inches, while two-year-old fish are twice this size and will weigh about 3 pounds. The youngsters feed heavily upon anchovies, but as they attain larger sizes they start feeding on other species. Stomachs of large individuals (100 pounds and bigger) have contained the remains of mackerel scad, Pacific mackerel, ocean whitefish, bonito, midshipmen, stingrays, small sharks, crabs, lobsters, mantis shrimp, and many other food items. A large giant sea bass has few problems of survival—mortality is caused by man, large sharks, parasites, disease, and old age.

Coastal Indians apparently caught fair numbers of giant sea bass during the several thousand years they inhabited southern California and offshore islands, but there is no fossil record of *Stereolepis*.

Fishery information.—During a "good" year sport fishermen, using hook and line and spears, probably catch 200 giant sea bass that weigh in excess of 100 pounds each. Perhaps 100 additional fish weighing 12 to 30 pounds each are taken by sport anglers, but catch records to support these figures are lacking. A 10- to 12-inch long Pacific mackerel makes a perfect bait for hook-and-line fishing as does a fillet of whitefish, a large herring or sardine, a lobster tail, and a few other items.

Although commercial landings ranged from 97,000 pounds (in 1922) to 861,000 (in 1934), very little of this was caught off California. Most of the catch is made by a few boats fishing with hook and line in Mexican waters for groupers, giant sea bass, and other market species that are fairly abundant between about Cape Colnett and Magdalena Bay. Beheaded and eviscerated fish are sold to the wholesalers who fillet and steak them for retail outlets. The flesh is mild but coarse-grained and "tough."

The sport fishery is regulated by a daily bag limit and other laws, so one should always check current restrictions before actively seeking out these trophy-sized giants.

Other family members.—Eight other sea basses are known from off California, but these are fairly easy to recognize by making a few simple observations and counts.

1. If the 2 dorsal fins are slightly separated, it is a striped bass.
2. If the third dorsal spine is very long and filamentous, it is a splittail bass.
3. If there are 11 dorsal spines, 10 dorsal rays, and 8 anal rays, it is a giant sea bass.
4. If there are 11 dorsal spines, 15 to 17 dorsal rays, 10 to 11 anal rays, and
   a. the end of the tail fin is jagged (serrate), it is a broomtail grouper.
   b. the end of the tail fin is smooth, it is a Gulf grouper.
5. If there are 10 dorsal spines, 13 to 15 dorsal rays, 7 to 8 anal rays, and
   a. the body is profusely covered with small brownish or black spots, it is a spotted sand bass.

   b. there are several vertical dusky bars on the sides, it is
      a barred sand bass.

   c. the upper body is mottled with pale yellow, whitish,
      or light brown blotches, it is a kelp bass.

6. If there are 10 dorsal spines, 17 dorsal rays, and 8 anal
   rays, it is a spotted cabrilla.

Meaning of name.—*Stereolepis* (firm scale) *gigas* (giant). A giant fish with firm scales.

Sparidae (Porgy Family)
Pacific Porgy
Calamus brachysomus (Lockington, 1880)

Distinguishing characters.—The Pacific porgy is the only fish in our waters with a perch-shaped (laterally compressed) body, a broad scaleless area between the lips and the eye, a pectoral fin that is longer than the head, and a brownish-pink body (darker above) with some silvery-white shining through. There is a smoothly rounded heavy bony ridge above each eye, and the rear margin of the preopercle is razor sharp. The outline of a Pacific porgy in side view is unmistakable. Juveniles have several chocolate-colored vertical bars on their sides, which are also present on adult porgies but fade almost as soon as the fish is hauled from the water.

Fig. 52.   *Calamus brachysomus*

[139]

Natural history notes.—Calamus brachysomus has been recorded from Oceanside (twice) to 150 miles beyond Lima, Peru, but it apparently does not spawn north of about Sebastian Viscaino Bay. Pacific porgies probably reach a length of slightly less than 2 feet and a weight of 5 pounds or more, but the largest one officially measured was 20 inches long. Its weight is unknown. An examination of more than a dozen sets of otoliths from large individuals indicates they attain an age of at least fifteen years.

Spawning apparently takes place during the spring months (based upon the presence of fish-of-the-year in mid-summer), and most individuals will spawn when they are three or four years old. We have no information on the number of eggs spawned per female, egg size, locality of spawning, or other facets of reproduction.

Young fish (up to 4 or 5 inches long) are abundant in very shallow water where the bottom is sandy or firm sandy mud, particularly in coastal bays and lagoons. The adults also live over smooth (non-rocky) bottom areas, but they usually are found in slightly greater depths. They appear to be most abundant in 20 to 60 feet of water, and there is one authentic capture from 225 feet. Their natural food seems to be an assortment of sedentary or semi-sedentary mollusks (clams and small snails) and crustaceans (shrimp and crabs). On rare occasions small fishes will be fed upon, but we have never observed more than one in a single stomach.

A large molar tooth (from the back part of the jaw) of *C. brachysomus* has been found in a Pleistocene deposit at San Pedro that was laid down more than 20,000 years ago when local ocean temperatures were much warmer than they are today.

Fishery information.—The Pacific porgy has been caught only twice north of Mexico, so it cannot be deemed a sought-after sport species in our waters. Porgies readily take a baited hook, but they are excellent bait thieves, and the large, flattened molarlike teeth in each jaw make it difficult to hook one solidly. They will take a variety of baits including cut anchovy, but best catches are made with pieces of clam, mussel, shrimp, or squid.

Porgies have been brought in from Mexican waters and sold as "tai" in the fresh fish markets, but landings have been sporadic and light. In 1931, landings of nearly 1 ton were reported, but this figure has never been equaled since then.

Other family members.—No other member of the family is known within several thousand miles of California.

Meaning of name.—*Calamus* (a quill or reed, for the quill-like interhaemal bone) *brachysomus* (short body).

Sphyraenidae (Barracuda Family)
California Barracuda
Sphyraena argentea Girard, 1854

Distinguishing characters.—No other fish in our waters could possibly be mistaken for the distinctively shaped barracuda. The two short-based dorsal fins are widely spaced, the quite large mouth is filled with a formidable array of teeth, and the tip of the pectoral fin fails to reach to beneath the front of the first dorsal.

Fig. 53. *Sphyraena argentea*

Natural history notes.—*Sphyraena argentea* has been caught from Prince William Sound, Alaska, to Magdalena Bay, Baja California, but its occurrence north of Point Conception is sporadic. It is a schooling species and seemingly prefers rather shallow waters close to shore, although schools often are encountered in the open ocean between some of our islands and the

[141]

mainland coast. Young barracuda often will enter coastal bays and lagoons.

The oldest fish encountered during two separate age studies was an eleven-year-old female that was 41 inches long and weighed 10 pounds. Individuals have been measured that were 46½ inches long, and a prize was awarded to an angler in 1963 for catching a California barracuda that reportedly weighed 18 pounds 3 ounces. Females grow faster and larger than males, and almost all fish heavier than 9 pounds are females. All three-year-old barracuda are sexually mature, but only three-fourths of the two-year-olds are capable of reproduction. The spawning season often extends from April through September, but peak spawning occurs in May, June, and July. A single female may release from 42,000 to 485,000 eggs per season depending upon her age and size. The pelagic eggs are 1.2 to 1.6 millimeters in diameter (.048 to .064 of an inch). Hatching apparently takes place in a matter of days after spawning.

Barracuda are voracious feeders and consume quantities of anchovies and other small prey species. In turn, barracuda are preyed upon by sea lions, porpoises, and giant sea bass, but large adults have few enemies other than man. Their viscera often are infested with encysted parasitic worms but these are probably not bothersome enough to be fatal to the hosting barracuda.

Otoliths of *S. argentea* have been found in several Pliocene and Pleistocene deposits in southern California, the oldest of which are estimated to have been laid down 12 million years ago. Otoliths, vertebrae, and other remains are sometimes abundant in coastal Indian middens from a few hundred to perhaps 7,000 years old.

Fishery information.—The California barracuda has been important to the state's sport and commercial fishermen for at least seventy-five years, but accurate catch statistics are not available for the earliest years. Sport anglers catch barracuda from jetties, breakwaters and docks, from skiffs, private boats, barges, and partyboats, and a few are even speared by skindivers, but catch figures are available only for partyboat fishermen. During the years 1936 through 1965 these fishermen caught between 87,000 (in 1956) and 1.2 million fish (in 1959) per year, but only twice during this period were fewer than 250,000 barracuda taken.

The commercial catch, made almost entirely with gill nets and by trolling (hook and line), is sold in the round (just as caught) for consumption fresh, although during some years a specialty pack (fish cakes) has been made by one or two canners, and a fair quantity is smoked. Since 1928, landings of fish caught off California have ranged from 50,000 pounds in 1956 (estimated 12,500 fish) to 4.4 million pounds in 1928 (1.1 million fish). Landings from Mexican waters have been about equal during most of these years.

In California, both the sport and commercial catch depend upon an influx of fish from the south when ocean temperatures warm up each summer. During periods when local temperatures have been warmer than usual the year around, barracuda have been caught in quantity off Monterey and have spawned successfully south off Point Conception.

Other family members.—S. argentea is the only member of the family known to California, although three other species are found between central Baja California and Peru.

Meaning of name.—Sphyraena (an ancient name pertaining to a hammer—in this case we do not see its applicability) *argentea* (silvery).

Stichaeidae (Prickleback Family)
Monkeyface Prickleback
Cebidichthys violaceus (Girard, 1854)

*Distinguishing characters.—*The monkeyface prickleback is the only member of the family in our waters that lacks pelvic fins and has soft rays in the posterior half of the dorsal fin. The gill membranes are free from the isthmus, or throat region, and there is but a single lateral line. Adults have a large fleshy lump on the top of the head.

Natural history notes.—*Cebidichthys violaceus* ranges from somewhere north of Crescent City, California, to about San Quintin Bay, Baja California, but it is not common south of Point Conception. They inhabit rocky areas from the intertidal (where they remain if there are pools of water when the tide is out) into depths of 80 feet or more. They are often abundant among the submerged rocks of breakwaters that are considerable distances from natural rocky habitat.

Fig. 54.   *Cebidichthys violaceus*

Few large monkeyface pricklebacks have been measured or weighed. There are reports of 3-footers having been caught, but 30 inches seems to be the most often quoted maximum length. Because these are such "round" measurements, we look upon them with a bit of skepticism. There is an authentic report of a 6-pound fish, and there are vague references to a 9-pounder, but this last figure cannot be associated with a specific fish. A 20-inch female speared at the Monterey breakwater in 1968 was between twelve and fifteen years old (weight not available), judged by the growth rings on its otoliths, so a 30- or 36-inch fish would either have to be a giant or a very old individual.

Spawning takes place in the spring, and it appears that most monkeyface pricklebacks will spawn when they are three or four years old. Their eggs are about a twelfth of an inch in diameter and are deposited in a protected spot on rocky substrate. They adhere to the rocks, and one of the parents (probably the female) guards them until they hatch. A ball of approximately 6,000 to 8,000 eggs was being tended by a monkeyface

[144]

prickleback estimated to be 30 inches long. No actual egg counts have been made, however.

We have no information on the length of time till hatching, or the habits and habitats of the larvae. Three-inch juveniles have been found in kelp beds and rocky intertidal areas.

An examination of several stomachs revealed an assortment of algae, but as with the opaleye, the animal life on the various seaweeds is probably the reason for its feeding on plant material. In an aquarium, monkeyface pricklebacks live happily and grow well on a diet of small shrimp, polychaete worms, and shucked clam or mussel. We do not know of any specific predator on monkeyface pricklebacks.

Otoliths of pricklebacks are abundant in coastal Indian middens north of Point Conception. Some of these are from *Cebidichthys violaceus*.

Fishery information.—A typical "blenny eel" fisherman uses a bamboo pole about 10 feet long and fishes cracks and crevices in and under rocks in surge channels and large pools left behind by a receding tide. At the small end of the pole he affixes a wire leader about 6 inches long and a medium-sized hook is warped to the end of this. The hook is baited with shrimp, mussel, clam, worm, or a small piece of salted herring, anchovy, or other baitfish, and the baited end is then poked into various submerged cracks, crevices, and holes that look as if they might hide a monkeyface prickleback. This technique is known as "poke-poling." A recent survey of the state's sport fisheries revealed that fewer than 1,000 monkeyface pricklebacks are caught each year by these pokepolers.

There is no commercial fishery for monkeyface pricklebacks.

Other family members.—Nine other pricklebacks are known from California, but only three of these ever exceed a foot in length. To our knowledge, no other publication differentiates the ten pricklebacks, so we have constructed a simple key that can be used to identify them. If any particular character in this key does not agree with the prickleback being identified, you must proceed either to the opposing couplet or to the next character if there is no opposing couplet.

1. No pelvic fins
    a. Last half of dorsal without needlelike spines, *Cebidichthys violaceus*
    b. Gill membranes attached to isthmus, *Anoplarchus purpurescens*
    c. Four lateral lines, both jaws of equal length
        i.   Dorsal begins above pectoral, *Phytichthys chirus*
        ii. Dorsal begins one eye diameter behind pectoral tip, *Xiphister mucosus*
        iii. Dorsal begins two or more eye diameters behind pectoral tip, *Xiphister atropurpureus*
2. Pelvic fins present
    a. Numerous parallel rows of well-defined pores on sides vertical to lateral line, *Plagiogrammus hopkinsii*
    b. A dense hairlike growth of cirri on top of head, *Chirolophis nugator*
    c. Two round black blotches on posterior end of dorsal fin, *Plectobranchus evides*
    d. Gill membranes not connected to each other or to isthmus posteriorly, *Lumpenus sagitta*
    e. Gill membranes joined to the isthmus, *Poroclinus rothrocki*

Meaning of name.—*Cebidichthys* (monkey fish) *violaceus* (violet).

Stromateidae (Butterfish Family)
Pacific Pompano
Peprilus simillimus (Ayres, 1860)
Distinguishing characters.—The metallic silvery laterally compressed (perchlike) body, lack of pelvic fins, deeply forked tail fin, and long pectorals will distinguish the Pacific pompano from all other species in our waters.

Natural history notes.—Peprilus simillimus ranges from British Columbia to about Abreojos Point, Baja California. It usually inhabits shallow water near shore, and often forms small, but fairly dense, schools. They are reported to reach a length of 10 inches, but few individuals have been seen that exceed 8. A female that was just over 8½ inches long weighed slightly less than 5 ounces, and appeared to be just past four years old.

Fig. 55.   *Peprilus simillimus*

Spawning takes place in the spring, extending perhaps into July, and the eggs are believed to be pelagic, but no valid information is available on the subject. They are believed to feed upon small crustaceans, but their food habits have not been studied. Small pompano have been found in the stomachs of California halibut, barracuda, and kelp bass; they are undoubtedly eaten by numerous other predators also.

To date their remains have not been found in Indian middens or fossil deposits, although *Peprilus* otoliths have been found in Atlantic coast Miocene. (Four species presently live off the Atlantic and Gulf coasts of the United States.)

Fishery information.—Pier fishermen catch fair numbers of Pacific pompano with snagging gear, and occasionally large individuals will be hooked on small pieces of clam or shrimp bait. Piers that are the most consistently productive are: Oceanside, Newport Beach, Huntington Beach, Belmont Shore (Long Beach), Malibu, and Cayucos.

The commercial catch is made almost entirely with encircling nets (purse seine, lampara, and bait net). The fish are delivered whole to the fresh fish markets where they usually bring an excellent price. They are highly prized by people of European extraction, and a few hundred pounds of freshly caught pompano seldom last more than half an hour at the waterfront wholesale fish markets. Of 62,000 pounds landed in 1965, two-thirds were from the Monterey area and most of the rest from Los Angeles.

Other family members.—No other member of the family occurs in our waters, although two or three species are known between about Magdalena Bay and Panama.

Meaning of name.—*Peprilus* (derived from the Greek, meaning one of Hesychian's unknown fish) *simillimus* (very similar — in reference to an Atlantic species).

Synodontidae (Lizardfish Family)
California Lizardfish
Synodus lucioceps (Ayres, 1855)

Distinguishing characters.—The typical body shape of the California lizardfish, combined with such characters as a broad head, an almost triangular snout, an adipose fin, tightly adhering scales on body and head, and numerous needle-sharp teeth in both jaws preclude mistaking this fish for any other species.

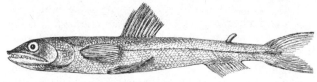

Fig. 56.   *Synodus lucioceps*

Natural history notes.—*Synodus lucioceps* has been recorded from San Francisco Bay to Cape San Lucas and throughout much of the Gulf of California. Larvae and young individuals up to about 3 inches are nearly transparent and have a series of black ventrolateral pigment spots beneath the skin (in the peritoneum) along the belly. These youngsters live in the upper water layers over deep water and often are found many miles offshore. The brownish-colored adults live at or near the bottom, usually in depths of 60 to 150 feet but sometimes shallower as well as deeper. The spawning season appears to be from about June through August, and at that time large individuals (15 inches long and longer) tend to aggregate or concentrate in sandy-bottom areas.

The largest S. *lucioceps* known was a 25 1/8-inch female weighing over 4 pounds that was caught in Santa Monica Bay in 1956. Its stomach contained the remains of a 7-inch white croaker and an anchovy that had been slightly shorter. Stomachs of other individuals that have contained food have almost always yielded one or more kinds of fish and an occasional squid.

The age of the 25 1/8-inch fish is not known, but otoliths of several individuals 20 to 22 inches long indicated they were eight or nine years old at that length. Lizardfish remains have been found in the stomachs of a few bottom-dwelling predators, and surface-feeding fishes take a toll of the larvae and early juveniles.

Synodus otoliths have been reported from the Eocene of Europe (possibly 60 million years old), but there is only one record of S. *lucioceps* as a fossil in California.

Fishery information.—In the southern Gulf of California and other tropic areas, lizardfish will swim off the bottom and take a fast-trolled spoon or plug that is as big or bigger than they are. Most captures in California are made with a baited hook fished near the bottom, however. Such catches are incidental to other fisheries, and we doubt that one fisherman in twenty-five who catches a lizardfish will take it home to eat. Purse-seine fishermen catch fair numbers of S. *lucioceps* each year

[149]

while fishing for other species in shallow water. There is no market for the species so they usually end up being saved as a curio, discarded, or in a cannery reduction pit where they are made into fish meal along with heads, tails, and viscera of the fish being canned.

The flesh is white but has a strong "fishy" odor when cooked and gives the impression of tasting a bit like iodine smells.

Other family members.—No other member of the family is known from our waters, although several other species occur in the lower Gulf of California and south of there to Chile.

Meaning of name.—*Synodus* (ancient name of a fish in which the teeth meet) *lucioceps* (pike head). A fish with a pikelike head in which the teeth meet.

Trichiuridae (Cutlassfish Family)
Pacific Cutlassfish
Trichiurus nitens Garman, 1899

Distinguishing characters.—The silver-colored band-shaped body that tapers to a hairlike tail will distinguish the Pacific cutlassfish from all other fishes in our waters. The dorsal fin commences just behind the eye and extends to the base of the hairlike tail, the anal fin consists of a series of tiny spinules which usually do not break the skin, and the mouth contains numerous long teeth which are barbed at their tips.

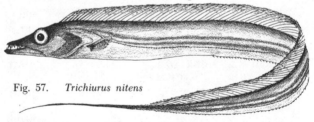

Fig. 57. *Trichiurus nitens*

Natural history notes.—*Trichiurus nitens* has been recorded from perhaps a half-dozen localities between San Pedro, California, and Paita, Peru. The first record from California (two fish) was in 1921. During the early 1930s cutlassfish were sometimes abundant in the Los Angeles Harbor area, but since 1934 they have been rarely seen and then only a few individuals at a time. The largest we have observed was a 44-inch long female that weighed about 3¼ pounds. This fish, netted during March 1962, was seven years old (judged by growth rings on its otoliths) and probably would have spawned in a month or two. A 33-inch male that was sucked onto the trash screens of a hydroelectric plant during February of the same year was four years old and in spawning condition. It would seem from these and other sporadic observations, that spawning in our area occurs during spring and early summer.

Most stomachs that have been examined have contained fish remains, primarily anchovies. The most anchovies found in a single stomach was four. We have found cutlassfish remains only in stomachs of tunas, but they undoubtedly are eaten by numerous other large predators including porpoises and sea lions.

Fishery information.—We have no record of a sport fisherman catching a cutlassfish, but commercial fishermen have fished for them with lampara nets, and have caught them incidentally during fishing operations for other species using bait nets and rockcod set lines. Only during the early 1930s were they taken in commercial quantities, and there was always a ready market for the few pounds to a ton or more that were landed almost daily for short periods. Related species are much sought after for food in many parts of the world, especially in the Orient.

Other family members.—Three other members of the family have been captured off California, but like the cutlassfish, none of these can be considered common in our waters. The four species are easily distinguishable upon casual observation. All have similar appearing heads, and all have very long bandlike bodies topped by lengthy dorsal fins. The characters that will separate them pertain to their caudal extremity, length, and mouth size.

1. If the body tapers to a hairlike tail it is a Pacific cutlassfish.
2. If the tail fin is mackerellike and:
   a. the body is about 15 times as long as deep, and
      i. the maxillary reaches to beneath the front of the eye and there are 78 to 81 dorsal fin rays, it is a black scabbardfish.
      ii. the maxillary reaches to beneath the center of the eye and there are 91 to 97 dorsal fin rays, it is a flathead scabbardfish.
   b. the body is about 25 times as long as deep, it is a razorback scabbardfish.

Meaning of name.—Trichiurus (little hair tail) *nitens* (shining, in reference to the silvery body).

Triglidae (Searobin Family)
Lumptail Searobin
Prionotus stephanophrys Lockington, 1880

*Distinguishing characters.—*The lumptail searobin is easily distinguished from all other marine fishes in our waters by the very long winglike pectoral fins (extending to above the sixth or seventh anal fin ray), which have the lowermost three rays loose and flexible. The head is quite large, and encased in bony armor, with spines and ridges which are best developed in young individuals. One or two vertebrae in the caudal peduncle region develop enlarged knobs in adults which are easily felt through the skin. It is these marble-sized lumps that give the fish its common name.

Natural history notes.—Prionotus stephanophrys ranges from off the Columbia River to the Gulf of California, and possibly to Peru. It is generally found over sandy or sandy-mud bottom areas, usually in depths of 60 to 150 feet, but it often occurs both shallower and deeper. In ancient times it was believed that searobins used their winglike pectorals to fly, but for many years now it has been known that they actually walk around

[152]

on the bottom on the three lowermost fingerlike pectoral rays. They will swim only if they wish to move at a more rapid rate.

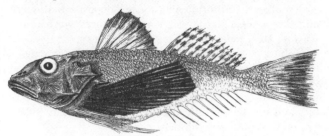

Fig. 58.   *Prionotus stephanophrys*

The largest lumptail searobin that we know of was caught off our coast. It was 15½ inches long and weighed just under 1 pound. This fish appeared to be eight or nine years old, as determined from growth rings on its otoliths.

Although the first lumptail searobin was caught off our coast in 1880, the species was not reported again until 1945, and even today fewer than two dozen individuals have been caught north of Mexico. Because of their rarity in our waters, and because no life-history studies have been conducted in Mexico where they are common, we have little information on the species.

Mature females (nearly ready to spawn) have been caught in Santa Monica Bay during June and July so some spawning takes place as late as July. A 1-inch long specimen, sent to California State Fisheries Laboratory in July 1957, had been found by a fisherman "inside a squid" that he had caught in a trawl net. Since it was not mutilated, it probably had not been eaten by the squid, but had become trapped in the mantle cavity while being banged around in the net. This tiny juvenile undoubtedly came from that year's spawning, so by applying logic one can assume a spring through early summer spawning season. The few stomachs we have

[153]

checked that had food in them contained mostly small shrimp.

Otoliths of *P. stephanophrys* have been found in several southern California Pleistocene deposits that were laid down when local water temperatures were much warmer than today.

Fishery information.—Sport fishermen have caught lumptail searobins on several occasions while fishing on the bottom with small hooks. Other captures have been made with gill nets and otter trawls. Searobins are edible, and on the Atlantic coast they are the basis for an important commercial fishery.

Other family members.—Although ten kinds of searobins live in the eastern tropical Pacific, only two have been captured off California. One of these, *Bellator xenisma,* has been taken only once (off Santa Barbara in 1958). Our two searobins can be easily distinguished by noting the shape of the snout (from above) or the pectoral fin length.

1. If the snout is evenly rounded and the pectoral reaches almost to the end of the anal fin it is a lumptail searobin.
2. If the snout is deeply notched and the pectoral fails to reach the anal fin it is a splitnose searobin.

Meaning of name.—*Prionotus* (saw back, in reference to three sawlike spines between the dorsal fins of some species) *stephanophrys* (crown eyebrow).

Xiphiidae (Swordfish Family)
Swordfish
Xiphias gladius Linnaeus, 1758
Distinguishing characters.—The flattened sword that forms the greatly prolonged upper jaw, the lack of pelvic fins, the scaleless body, and the wide keels on each side of the base of the tail are sufficient to distinguish the swordfish from all other species.

Natural history notes.—*Xiphias gladius* ranges throughout all warm temperate seas of the world. On our coast they are most plentiful seasonally (June through September) south of Point Conception, but during years when the ocean is warm farther north they may wander as far as Oregon. They are not schooling fishes, but there is a tendency for individuals to aggregate over fairly broad expanses of the ocean. They usually are seen lazing along at the surface, but with the advent of deep submergence vehicles, swordfish have been seen and photographed more than 2,000 feet down, particularly at the heads of submarine canyons. Some of these swordfish have fed upon squid and other creatures attracted to the lights of the "diving saucers," and one individual attacked the Woods Hole Oceanographic Institute's *Alvin* while at a depth of 2,000 feet. In this case the swordfish came out a loser, getting its sword wedged so tightly into a seam on *Alvin* that it could not extricate itself.

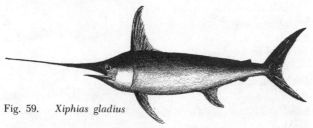

Fig. 59.   *Xiphias gladius*

The currently recognized world record swordfish weighed 1,182 pounds and was 14 feet 11¼ inches long from tip to tip. It was caught with rod and reel at Iquique, Chile, in 1953. A 6-foot long *Xiphias* harpooned near Santa Catalina Island in 1958 weighed 38 pounds. It is the smallest known from our coast. That same year a ripe female was harpooned off Santa Catalina Island. Its enlarged ovaries were estimated to contain 50 million eggs. There is no record of a spawning ground closer to California than the Marquesas Islands.

In the Mediterranean, where *Xiphias* spawning has been studied, some females will liberate eggs during every month of the year, but peak activity occurs during

[155]

June and July. The eggs are about 1.8 millimeters in diameter (.072 of an inch), and they require about two and one-half days to hatch. No information is available on swordfish age and growth, but indirect evidence leads us to believe that they probably grow quite rapidly and do not live for a great number of years.

An examination of numerous stomachs from fish harpooned in our commercial fishery revealed that most swordfish had fed on anchovies and cephalopods (presumably squid), but that they depended heavily upon such other fishes as hake, jack mackerel, and shortbelly rockcod, too. A few stomachs contained only deep-sea fishes: lanternfishes, barracudinas, pencil smelt, and other oddities. Man and large sharks probably are the only enemies that make inroads on the adult swordfish population.

Swords and jaw fragments of *X. gladius* have been found in coastal Indian middens, and an extinct swordfish (not *X. gladius*) left remains in Miocene and older deposits throughout the United States.

Fishery information.—The time and expense involved in sport fishing for swordfish in our waters places them beyond the reach of the average angler. The total California sport catch probably averages fewer than five fish per year.

Prior to the mid-1920s there was practically no market for swordfish, but since about 1927 it has been a choice consumer item and consistently pays the fisherman one of the highest per-pound prices he receives. Commercially they are taken by harpooning, and the catch is delivered to the fresh fish markets in a beheaded and eviscerated condition with all fins removed. The catch has fluctuated widely since 1927, depending primarily on how warm the ocean gets off our coast and how good the albacore season is. In 1950 and 1954 landings were the poorest on record as only 26,000 and 23,000 pounds respectively were landed. The best catch since the 1927 awakening of the fishery was 1.1 million pounds in 1948. Most of the catch is made around southern California's offshore islands.

The finding of high levels of mercury in swordfish flesh late in 1970, and subsequent warnings by various public health agencies that such levels are higher than should be tolerated for human consumption, probably will destroy the market for these tasty fish. Although proof is lacking, we believe that these levels of mercury have been in swordfish since *Xiphias gladius* first swam and fed in the world's oceans, millions of years before man started polluting the environment.

Other family members.—There is no other member of the family.

Meaning of name.—*Xiphias* (the ancient name for the fish) *gladius* (sword).

Zaproridae (Prowfish Family)
Prowfish
Zaprora silenus Jordan, 1896

Distinguishing characters.—The prowfish is best distinguished by its typical elongate body shape and the numerous large pores on the sides and top of the head. Additional distinctive characters include the lack of ventral fins, an elongate dorsal fin, and an anal fin that is only about half as long as the dorsal. The body color is variable but usually includes numerous yellowish spots along the sides and some turquoise blue around the head pores.

Fig. 60.   *Zaprora silenus*

[157]

Natural history notes.—Zaprora silenus ranges from the Aleutian Islands, to Bodega Bay, California. It is a rare fish throughout its range, with most adults being captured in bottom trawling gear being fished at depths shallower than 600 feet. There is one record of a prowfish from nearly 1,200 feet, however.

The largest individual noted in the literature appears to be a fish that was 34½ inches long. We have no additional statistics on this specimen, but a 33-inch male trawled from 330 feet near Eureka in April 1968 weighed 15½ pounds. Growth zones on the otoliths of this fish indicated it was twelve years old and had first spawned when it was four. We have no information on spawning season, number of eggs per female, or other facets of reproduction. Larvae up to 3 inches long have been captured with plankton nets towed at the surface, so the species does spend part of its first year in the pelagic environment.

We have not observed any food in the stomachs of two individuals, but jellyfish remains have been reported by other workers. We do not feel that jellyfish provide the bulk of their diet, however. No specific predator on adult prowfish is known.

*Fishery information.—*There is no sport or commercial fishery for prowfish. Adults are caught primarily in otter trawls being fished for other species, but they are quite rare, as are the larvae. During the 1930s, the International Halibut Commission captured 36 larvae in plankton nets, but they made more than 2,500 tows in catching these three dozen fish. The flesh is reported to range in color from white to almost red, but is mild and tasty.

*Other family members.—*No other member of the family is known.

Meaning of name.—Zaprora (an intensive particle, and prow, with reference to the shape of the forehead) *silenus* (a drunken demigod, covered with slime, in allusion to the large mucous pores on the head).

APPENDIX I

Checklist of Marine Food
and Game Fishes of California

(Asterisk indicates species description is given in text.)

Acipenseridae
  Acipenser medirostris Ayres,
    1854 — Green sturgeon°

  Acipenser transmontanus
    Richardson, 1836 — White sturgeon
Albulidae
  Albula vulpes (Linnaeus, 1758) — Bonefish°
Anarhichadidae
  Anarrhichthys ocellatus Ayres,
    1855 — Wolf-eel°
Anoplopomatidae
  Anoplopoma fimbria (Pallas,
    1814) — Sablefish°

  Erilepis zonifer (Lockington,
    1880) — Skilfish
Ariidae
  Bagre panamensis (Gill, 1863) — Chihuil°
Atherinidae
  Atherinops affinis (Ayres, 1860) — Topsmelt

  Atherinopsis californiensis Girard,
    1854 — Jacksmelt

  Leuresthes tenuis (Ayres, 1860) — California grunion°
Balistidae
  Balistes polylepis Steindachner,
    1876 — Finescale triggerfish°

  Xanthichthys mento (Jordan and
    Gilbert, 1882) — Redtail triggerfish
Batrachoididae
  Porichthys myriaster Hubbs &
    Schultz, 1939 — Specklefin midshipman

  Porichthys notatus Girard, 1854 — Plainfin midshipman°
Belonidae
  Strongylura exilis (Girard, 1854) — California needlefish°
Bothidae
  Citharichthys sordidus (Girard,
    1854) — Pacific sanddab°

Citharichthys stigmaeus Jordan &
  Gilbert, 1882   Speckled sanddab
Citharichthys xanthostigma
  Gilbert, 1890   Longfin sanddab
Hippoglossina stomata Eigenmann
  & Eigenmann, 1890   Bigmouth sole
Paralichthys californicus (Ayres,
  1859)   California halibut°
Xystreurys liolepis Jordan &
  Gilbert, 1880   Fantail sole
Bramidae
  Brama japonica Hilgendorf, 1878   Pacific pomfret°
  Pteraclis velifera (Pallas, 1770)   Fanfish
  Taractes longipinnis (Lowe, 1843)   Bigscale pomfret
Branchiostegidae
  Caulolatilus princeps (Jenyns,
    1842)   Ocean whitefish°
Carangidae
  Caranx caballus Günther, 1868   Green jack
  Chloroscombrus orqueta Jordan
    & Gilbert, 1883   Pacific bumper
  Decapterus hypodus Gill, 1862   Mexican scad
  Naucrates ductor (Linnaeus,
    1758)   Pilotfish
  Nematistius pectoralis Gill, 1862   Roosterfish
  Oligoplites saurus (Bloch &
    Schneider, 1801)   Leatherjacket
  Seriola colburni Evermann &
    Clark, 1928   Pacific amberjack
  Seriola dorsalis (Gill, 1863)   Yellowtail°
  Trachinotus paitensis Cuvier,
    1832   Paloma pompano
  Trachinotus rhodopus (Gill, 1863)   Gafftopsail pompano
  Trachurus symmetricus (Ayres,
    1855)   Jack mackerel°
  Vomer declivifrons Meek &
    Hildebrand, 1925   Pacific moonfish
Clupeidae
  Alosa sapidissima (Wilson, 1811)   American shad
  Clupea pallasii Valenciennes,
    1847   Pacific herring
  Dorosoma petenense (Günther,
    1867)   Threadfin shad

Etrumeus teres (DeKay, 1842)	Round herring
Harengula thrissina (Jordan & Gilbert, 1882)	Flatiron herring
Opisthonema medirastre Berry & Barrett, 1963	Middling thread herring
Sardinops caeruleus (Girard, 1854)	Pacific sardine°
Coryphaenidae	
Coryphaena hippurus Linnaeus, 1758	Common dolphin°
Cottidae	
Artedius corallinus (Hubbs, 1926)	Coralline sculpin
Artedius creaseri (Hubbs, 1926)	Roughcheek sculpin
Artedius fenestralis Jordan & Gilbert, 1883	Padded sculpin
Artedius harringtoni (Starks, 1896)	Scalyhead sculpin
Artedius lateralis (Girard, 1854)	Smoothhead sculpin
Artedius notospilotus Girard, 1856	Bonehead sculpin
Ascelichthys rhodorus Jordan & Gilbert, 1880	Rosylip sculpin
Asemichthys vinculus (Bolin, 1950)	Smoothgum sculpin
Blepsias cirrhosus (Pallas, 1814)	Silverspotted sculpin
Chitonotus pugetensis (Steindachner, 1876)	Roughback sculpin
Clinocottus acuticeps (Gilbert, 1895)	Sharpnose sculpin
Clinocottus analis (Girard, 1858)	Wooly sculpin
Clinocottus embryum (Jordan & Starks, 1895)	Calico sculpin
Clinocottus globiceps (Girard, 1858)	Mosshead sculpin
Clinocottus recalvus (Greeley, 1899)	Bald sculpin
Enophrys bison (Girard, 1854)	Buffalo sculpin
Enophrys taurina Gilbert, 1914	Bull sculpin
Hemilepidotus hemilepidotus (Tilesius, 1810)	Red Irish lord
Hemilepidotus spinosus (Ayres, 1855)	Brown Irish lord

Icelinus burchami Evermann & Goldsborough, 1907	Dusky sculpin
Icelinus cavifrons Gilbert, 1890	Pit-head sculpin
Icelinus filamentosus Gilbert, 1890	Threadfin sculpin
Icelinus fimbriatus Gilbert, 1890	Fringed sculpin
Icelinus oculatus Gilbert, 1890	Frogmouth sculpin
Icelinus quadriseriatus (Lockington, 1880)	Yellowchin sculpin
Icelinus tenuis Gilbert, 1890	Spotfin sculpin
Jordania zonope Starks, 1896	Longfin sculpin
Leiocottus hirundo Girard, 1856	Lavender sculpin
Leptocottus armatus Girard, 1854	Pacific staghorn sculpin
Nautichthys oculofasciatus (Girard, 1858)	Sailfin sculpin
Oligocottus maculosus Girard, 1856	Tidepool sculpin
Oligocottus rimensis (Greeley, 1899)	Saddleback sculpin
Oligocottus rubellio (Greeley, 1899)	Rosy sculpin
Oligocottus snyderi Greeley, 1898	Fluffy sculpin
Orthonopias triacis Starks & Mann, 1911	Snubnose sculpin
Paricelinus hopliticus Eigenmann & Eigenmann, 1889	Thornback sculpin
Radulinus asprellus Gilbert, 1890	Slim sculpin
Radulinus boleoides Gilbert, 1898	Darter sculpin
Rhamphocottus richardsoni Günther, 1874	Grunt sculpin
Scorpaenichthys marmoratus Girard, 1854	Cabezon°
Synchirus gilli Bean, 1889	Manacled sculpin
Zesticelus profundorum (Gilbert, 1895)	Flabby sculpin

Embiotocidae

Amphistichus argenteus Agassiz, 1854	Barred surfperch°
Amphistichus koelzi (Hubbs, 1933)	Calico surfperch

Amphistichus rhodoterus (Agassiz, 1854) — Redtail surfperch

Brachyistius frenatus Gill, 1862 — Kelp perch

Cymatogaster aggregata Gibbons, 1854 — Shiner perch

Cymatogaster gracilis Tarp, 1952 — Island seaperch

Damalichthys vacca Girard, 1855 — Pile perch

Embiotoca jacksoni Agassiz, 1853 — Black perch

Embiotoca lateralis Agassiz, 1854 — Striped seaperch

Hyperprosopon anale Agassiz, 1861 — Spotfin surfperch

Hyperprosopon argenteum Gibbons, 1854 — Walleye surfperch

Hyperprosopon ellipticum (Gibbons, 1854) — Silver surfperch

Echeneidae

Phtheirichthys lineatus (Menzies, 1791) — Slender suckerfish

Remora australis (Bennett, 1840) — Whalesucker

Remora brachyptera (Lowe, 1839) — Spearfish remora

Remora remora (Linnaeus, 1758) — Common remora°

Remorina albescens (Temminck & Schlegel, 1850) — White suckerfish

Rhombochirus osteochir (Cuvier, 1829) — Marlinsucker

Hypsurus caryi (Agassiz, 1853) — Rainbow seaperch

Micrometrus aurora (Jordan & Gilbert, 1880) — Reef perch

Micrometrus minimus (Gibbons, 1854) — Dwarf perch

Phanerodon atripes (Jordan & Gilbert, 1880) — Sharpnose seaperch

Phanerodon furcatus Girard, 1854 — White seaperch

Rhacochilus toxotes Agassiz, 1854 — Rubberlip seaperch°

Zalembius rosaceus (Jordan & Gilbert, 1880) — Pink seaperch

Engraulidae

Anchoa compressa (Girard, 1858) — Deepbody anchovy

Anchoa delicatissima (Girard, 1854) — Slough anchovy

[163]

Anchoviella miarcha (Jordan & Gilbert, 1881)	Slim anchovy
Cetengraulis mysticetus (Günther, 1866)	Anchoveta
Engraulis mordax Girard, 1854	Northern anchovy°
Exocoetidae	
Cypselurus californicus (Cooper, 1863)	California flyingfish°
Cypselurus heterurus (Rafinesque, 1810)	Blotchwing flyingfish
Prognichthys rondeletii (Valenciennes, 1846)	Blackwing flyingfish
Gadidae	
Gadus macrocephalus Tilesius, 1810	Pacific cod
Microgadus proximus (Girard, 1854)	Pacific tomcod°
Theragra chalcogramma (Pallas, 1814)	Walleye pollock
Hexagrammidae	
Hexagrammos decagrammus (Pallas, 1810)	Kelp greenling°
Hexagrammos lagocephalus (Pallas, 1810)	Rock greenling
Ophiodon elongatus Girard, 1854	Lingcod°
Oxylebius pictus Gill, 1862	Convictfish
Pleurogrammus monopterygius (Pallas, 1810)	Atka mackerel
Icosteidae	
Icosteus aenigmaticus Lockington, 1880	Ragfish°
Istiophoridae	
Istiompax indica (Cuvier, 1831)	Black marlin
Istiophorus platypterus (Shaw & Nodder, 1792)	Pacific sailfish
Tetrapturus angustirostris Tanaka, 1915	Shortbill spearfish
Tetrapturus audax (Philippi, 1887)	Striped marlin°
Kyphosidae	
Girella nigricans (Ayres, 1860)	Opaleye°
Hermosilla azurea Jenkins & Evermann, 1888	Zebraperch

Medialuna californiensis
(Steindachner, 1875) — Halfmoon

Labridae
Halichoeres semicinctus (Ayres,
1859) — Rock wrasse

Oxyjulis californica (Günther,
1861) — Señorita

Pimelometopon pulchrum (Ayres,
1854) — California sheephead°

Lampridae
Lampris regius (Bonnaterre, 1788) — Opah°

Luvaridae
Luvarus imperialis Rafinesque,
1810 — Louvar°

Merlucciidae
Merluccius productus (Ayres,
1855) — Pacific hake°

Molidae
Mola mola (Linnaeus, 1758) — Ocean sunfish°
Ranzania laevis (Pennant, 1776) — Slender mola

Mugilidae
Mugil cephalus Linnaeus, 1758 — Striped mullet°

Muraenidae
Gymnothorax mordax (Ayres,
1859) — California moray°

Osmeridae
Allosmerus elongatus (Ayres,
1854) — Whitebait smelt

Hypomesus pretiosus (Girard,
1854) — Surf smelt

Hypomesus transpacificus
McAllister, 1963 — Delta smelt

Spirinchus starksi (Fisk, 1913) — Night smelt°

Spirinchus thaleichthys (Ayres,
1860) — Longfin smelt

Thaleichthys pacificus
(Richardson, 1836) — Eulachon

Pleuronectidae
Atheresthes stomias (Jordan &
Gilbert, 1880) — Arrowtooth flounder

Embassichthys bathybius
(Gilbert, 1890) — Deepsea sole

Eopsetta jordani (Lockington, 1879)	Petrale sole
Glyptocephalus zachirus Lockington, 1879	Rex sole
Hippoglossoides elassodon Jordan & Gilbert, 1880	Flathead sole
Hippoglossus stenolepis Schmidt, 1904	Pacific halibut
Hypsopsetta guttulata (Girard, 1856)	Diamond turbot
Isopsetta isolepis (Lockington, 1880)	Butter sole
Lepidopsetta bilineata (Ayres, 1855)	Rock sole
Lyopsetta exilis (Jordan & Gilbert, 1880)	Slender sole
Microstomus pacificus (Lockington, 1879)	Dover sole°
Parophrys vetulus Girard, 1854	English sole
Platichthys stellatus (Pallas, 1814)	Starry flounder°
Pleuronichthys coenosus Girard, 1854	C-O turbot
Pleuronichthys decurrens Jordan & Gilbert, 1881	Curlfin turbot
Pleuronichthys ritteri Starks & Morris, 1907	Spotted turbot
Pleuronichthys verticalis Jordan & Gilbert, 1880	Hornyhead turbot
Psettichthys melanostictus Girard, 1854	Sand sole
Reinhardtius hippoglossoides (Walbaum, 1792)	Greenland halibut
Polynemidae	
Polydactylus approximans (Lay & Bennett, 1839)	Blue bobo°
Polydactylus opercularis (Gill, 1863)	Yellow bobo
Pomacentridae	
Chromis punctipinnis (Cooper, 1863)	Blacksmith°
Hypsypops rubicundus (Girard, 1854)	Garibaldi

Pomadasyidae
  Anisotremus davidsonii
    (Steindachner, 1875)   Sargo°

  Xenistius californiensis
    (Steindachner, 1875)   Salema
Salmonidae
  Oncorhynchus gorbuscha
    (Walbaum, 1792)   Pink salmon

  Oncorhynchus keta (Walbaum,
    1792)   Chum salmon

  Oncorhynchus kisutch (Walbaum,
    1792)   Silver or Coho salmon

  Oncorhynchus nerka (Walbaum,
    1792)   Sockeye salmon°

  Oncorhynchus tshawytscha
    (Walbaum, 1792)   King salmon°

  Salmo gairdnerii Richardson,
    1836   Rainbow trout or "steelhead"
Sciaenidae
  Cheilotrema saturnum (Girard,
    1858)   Black croaker

  Cynoscion nobilis (Ayres, 1860)   White seabass°

  Cynoscion parvipinnis Ayres,
    1861   Shortfin corvina

  Genyonemus lineatus (Ayres,
    1855)   White croaker

  Menticirrhus undulatus (Girard,
    1854)   California corbina°

  Roncador stearnsii (Steindachner,
    1875)   Spotfin croaker

  Seriphus politus Ayres, 1860   Queenfish

  Umbrina roncador Jordan &
    Gilbert, 1882   Yellowfin croaker
Scomberesocidae
  Cololabis saira (Brevoort, 1856)   Pacific saury°
Scombridae
  Allothunnus fallai Serventy, 1948   Slender tuna

  Auxis rochei (Risso, 1810)   Bullet mackerel

  Auxis thazard Lacépède, 1802   Frigate mackerel

  Euthynnus lineatus Kishinouye,
    1920   Black skipjack

  Euthynnus yaito Kishinouye,
    1915   Kawa Kawa

Katsuwonus pelamis (Linnaeus, 1758)	Oceanic skipjack
Lepidocybium flavobrunneum (Smith, 1849)	Escolar
Sarda lineolata (Girard, 1858)	California bonito
Scomber japonicus Houttuyn, 1782	Pacific mackerel°
Scomberomorus concolor (Lockington, 1879)	Monterey Spanish mackerel
Scomberomorus sierra Jordan & Starks, 1895	Sierra
Thunnus alalunga (Bonnaterre, 1788)	Albacore°
Thunnus albacares (Bonnaterre, 1788)	Yellowfin tuna
Thunnus obesus (Lowe, 1839)	Bigeye tuna
Thunnus thynnus (Linnaeus, 1758)	Bluefin tuna
Scorpaenidae	
Scorpaena guttata Girard, 1854	California scorpionfish
Scorpaenodes xyris (Jordan & Gilbert, 1882)	Rainbow scorpionfish
Sebastes aleutianus (Jordan & Evermann, 1898)	Blackthroat rockcod
Sebastes alutus (Gilbert, 1890)	Pacific ocean perch
Sebastes atrovirens (Jordan & Gilbert, 1880)	Kelp rockcod
Sebastes auriculatus Girard, 1854	Brown rockcod
Sebastes aurora (Gilbert, 1890)	Aurora rockcod
Sebastes babcocki (Thompson, 1914)	Redbanded rockcod
Sebastes brevispinis (Bean, 1884)	Silvergray rockcod
Sebastes carnatus (Jordan & Gilbert, 1880)	Gopher rockcod
Sebastes caurinus Richardson, 1845	Copper rockcod
Sebastes chlorostictus (Jordan & Gilbert, 1880)	Greenspotted rockcod
Sebastes chrysomelas (Jordan & Gilbert, 1880)	Black-and-yellow rockcod
Sebastes constellatus (Jordan & Gilbert, 1880)	Starry rockcod

Sebastes crameri (Jordan, 1896)	Darkblotched rockcod
Sebastes dallii (Eigenmann & Beeson, 1894)	Calico rockcod
Sebastes diploproa (Gilbert, 1890)	Splitnose rockcod
Sebastes elongatus Ayres, 1859	Greenstriped rockcod
Sebastes entomelas (Jordan & Gilbert, 1880)	Widow rockcod
Sebastes eos (Eigenmann & Eigenmann, 1890)	Pink rockcod
Sebastes flavidus (Ayres, 1862)	Yellowtail rockcod
Sebastes gilli (Eigenmann, 1891)	Bronzespotted rockcod
Sebastes goodei (Eigenmann & Eigenmann, 1890)	Chilipepper
Sebastes helvomaculatus Ayres, 1859	Rosethorn rockcod
Sebastes hopkinsi (Cramer, 1895)	Squarespot rockcod
Sebastes jordani (Gilbert, 1893)	Shortbelly rockcod
Sebastes levis (Eigenmann & Eigenmann, 1889)	Cowcod°
Sebastes macdonaldi (Eigenmann & Beeson, 1893)	Mexican rockcod
Sebastes maliger (Jordan & Gilbert, 1880)	Quillback rockcod
Sebastes melanops Girard, 1856	Black rockcod
Sebastes melanostomus (Eigenmann & Eigenmann, 1890)	Blackgill rockcod
Sebastes miniatus (Jordan & Gilbert, 1880)	Vermilion rockcod
Sebastes mystinus (Jordan & Gilbert, 1880)	Blue rockcod
Sebastes nebulosus Ayres, 1854	China rockcod
Sebastes nigrocinctus Ayres, 1859	Tiger rockcod
Sebastes ovalis (Ayres, 1862)	Speckled rockcod
Sebastes paucispinis Ayres, 1854	Bocaccio°
Sebastes phillipsi (Fitch, 1964)	Chameleon rockcod
Sebastes pinniger (Gill, 1864)	Canary rockcod
Sebastes proriger (Jordan & Gilbert, 1880)	Redstripe rockcod
Sebastes rastrelliger (Jordan & Gilbert, 1880)	Grass rockcod

Sebastes reedi (Westrheim & Tsuyuki, 1967)   Yellowmouth rockcod

Sebastes rhodochloris (Jordan & Gilbert, 1880)   Swordspine rockcod

Sebastes rosaceus Girard, 1854   Rosy rockcod

Sebastes ruberrimus (Cramer, 1895)   Yelloweye rockcod

Sebastes rubrivinctus (Jordan & Gilbert, 1880)   Flag rockcod

Sebastes rufus (Eigenmann & Eigenmann, 1890)   Bank rockcod

Sebastes saxicola (Gilbert, 1890)   Stripetail rockcod

Sebastes semicinctus (Gilbert, 1896)   Halfbanded rockcod

Sebastes serranoides (Eigenmann & Eigenmann, 1890)   Olive rockcod

Sebastes serriceps (Jordan & Gilbert, 1880)   Treefish

Sebastes umbrosus (Jordan & Gilbert, 1882)   Honeycomb rockcod

Sebastes vexillaris (Jordan & Gilbert, 1880)   Whitebelly rockcod

Sebastes wilsoni (Gilbert, 1915)   Pygmy rockcod

Sebastes zacentrus (Gilbert, 1890)   Sharpchin rockcod

Sebastolobus alascanus Bean, 1890   Shortspine thornyhead

Sebastolobus altivelis Gilbert, 1895   Longspine thornyhead

Serranidae

Epinephelus analogus Gill, 1863   Spotted Cabrilla

Hemanthias peruanus (Steindachner, 1875)   Splittail bass

Mycteroperca jordani (Jenkins & Evermann, 1889)   Gulf grouper

Mycteroperca xenarcha Jordan, 1887   Broomtail grouper

Paralabrax clathratus (Girard, 1854)   Kelp bass

Paralabrax maculatofasciatus (Steindachner, 1868)   Spotted sand bass

Paralabrax nebulifer (Girard, 1854)   Barred sand bass

Morone saxatilis (Walbaum, 1792)   Striped bass
Stereolepis gigas Ayres, 1859   Giant sea bass°
Sparidae
Calamus brachysomus (Lockington, 1880)   Pacific porgy°
Sphyraenidae
Sphyraena argentea Girard, 1854   California barracuda°
Stichaeidae
Anoplarchus purpurescens Gill, 1861   High cockscomb
Cebidichthys violaceus (Girard, 1854)   Monkeyface prickleback°
Chirolophis nugator (Jordan & Williams, 1895)   Mosshead warbonnet
Lumpenus sagitta Wilimovsky, 1956   Snake prickleback
Phytichthys chirus (Jordan & Gilbert, 1880)   Ribbon prickleback
Plagiogrammus hopkinsii Bean, 1893   Crisscross prickleback
Plectobranchus evides Gilbert, 1890   Bluebarred prickleback
Poroclinus rothrocki Bean, 1890   Whitebarred prickleback
Xiphister atropurpureus (Kittlitz, 1858)   Black prickleback
Xiphister mucosus (Girard, 1858)   Rock prickleback
Stromateidae
Peprilus simillimus (Ayres, 1860)   Pacific pompano°
Synodontidae
Synodus lucioceps (Ayres, 1855)   California lizardfish°
Trichiuridae
Aphanopus carbo Lowe, 1839   Flathead scabbardfish
Assurger anzac (Alexander, 1916)   Razorback scabbardfish
Lepidopus xantusi Goode & Bean, 1896   Black scabbardfish
Trichiurus nitens Garman, 1899   Pacific cutlassfish°
Triglidae
Bellator xenisma (Jordan & Bollman, 1890)   Splitnose searobin°
Prionotus stephanophrys Lockington, 1880   Lumptail searobin°
Xiphiidae
Xiphias gladius Linnaeus, 1758   Swordfish°
Zaproridae
Zaprora silenus Jordan, 1896   Prowfish°

APPENDIX II

HELPFUL REFERENCES

Baxter, John L. 1966. Inshore fishes of California. Calif. Dept. Fish & Game, Sacramento. 80 pp.

Bolin, Rolf L. 1944. A review of the marine cottid fishes of California. Stanford Ich. Bull. 3(1): 1-135.

Cannon, Raymond. 1964. How to fish the Pacific Coast. 2nd edition. Lane Publ. Co., Menlo Park. 337 pp.

Clemens, W. A., and G. V. Wilby. 1961. Fishes of the Pacific Coast of Canada. 2nd edition. Fish. Res. Bd. Canada, Bull. 68, 443 pp.

Fitch, John E. 1969. Offshore fishes of California. 4th revision. Calif. Dept. Fish & Game, Sacramento. 80 pp.

Gabrielson, Ira N. 1963. The fisherman's encyclopedia. 2nd edition. The Stackpole Co., Harrisburg, Pa. 759 pp.

Grosvenor, Melville Bell (ed.). 1965. Wondrous world of fishes. National Geographic Society, Wash., D.C. 367 pp.

Herald, Earl S. 1961. Living fishes of the world. Doubleday & Co., New York. 304 pp.

International Game Fish Association. 1969. World record marine fishes. Int. Game Fish Assoc., Fort Lauderdale, Fla. 16 pp. (A new revision is issued each year.)

McClane, A. J. 1965. McClane's standard fishing encyclopedia and international angling guide. Holt, Rinehart & Winston, New York. 1057 pp.

Miller, Daniel J., Dan Gotshall, and Richard Nitsos. 1965. A field guide to some common ocean sport fishes of California. 2nd revision. Calif. Dept. Fish & Game, Sacramento. 87 pp.

Phillips, Julius B. 1957. A review of the rockfishes of California (family Scorpaenidae). Calif. Dept. Fish & Game, Fish Bull., 104:1-158.

Roedal, Phil M. 1953. Common ocean fishes of the California coast. Calif. Dept. Fish & Game, Fish Bull., 91:1-184.

Zim, Herbert S., and H. H. Shoemaker. 1956. Fishes. A guide to familiar American species. Simon & Schuster, New York. 160 pp.

GLOSSARY

amphipods: group name for small shrimplike or crablike crustaceans.

bait net: an encircling net with a centrally located bag of fine mesh for holding fish while the net is being retrieved; operated similarly to a purse seine but is not pursed.

barbel: a slender fleshy chin whisker, typical of the cods and their relatives, but also found in a few other food and game fishes.

beach seine: a wall-like net of fine mesh that can be set parallel to shore and then pulled to the beach from both ends so as to catch the entrapped shallow-water fishes.

cephalopods: group name for octopi and squids.

cirri: filaments or fleshy appendages, usually found in the head region.

copepods: group name for small flealike or shrimplike crustaceans that abound in much of the ocean.

crustaceans: group name for a class of arthropods (joint-legged animals with horny or chitinous exoskeletons) which includes such creatures as lobsters, crabs, shrimp, amphipods, euphausiids, and others.

elasmobranchs: fishes having cartilaginous skeletons, namely sharks, skates, and rays.

euphausiids group name for krill or shrimplike creatures fed upon by many marine fishes.

fecundity: the egg-bearing capacity of female fishes, differing greatly depending upon species and size (age).

fish-of-the-year: fish hatched during current year.

gill net: any net that hangs wall-like in the water and catches fish that are too large to swim through it; the fish usually penetrates a mesh or opening as far as its gills and then cannot back out.

gill rakers: toothlike bony structures along the anterior edge of a gill arch, forming a comblike straining device in many fishes.

hermaphrodite: a fish that has both male and female organs, a normal arrangement in some species.

Indian midden: refuse heap left by Indians, usually marking camp sites.

invertebrates: animals lacking backbones.

isthmus: the fleshy interspace beneath the head and between the gill openings.

laterally compressed: perchlike in that it is flattened from side to side.

leptocephalus: a transparent elongate or leaf-shaped larval stage that occurs in bonefishes, eels, and some related families.

mouth:

inferior —on the underside of the fish.

oblique —at an angle of 45° or greater when closed.

subterminal —slightly overhung by snout.

terminal —upper and lower jaws meet to form anteriormost part of head.

mysids: group name for shrimplike creatures that abound throughout the world oceans.

otolith: a calcareous concretion in the inner ear of a fish, functioning as organs of hearing and balance.

otter trawls: tapering, sacklike nets which are lowered to the bottom where they catch fish as they are towed by a boat on the surface.

parr: dusky oval blotches on sides of young salmon or trout.

partyboat: a fishing boat that operates on a regular schedule and sells fishing space, on a "first-come, first-serve" basis, as opposed to charter boats.

pelagic: living in open waters in contrast to bottom or inshore.

pharyngeal teeth: patches of teeth in the throat region of a fish, lying behind the gills and in front of the esophagus.

photophores: complicated organs for emitting light or luminescing; usually found on sides, heads, and bellies of deep-sea fishes but may be elsewhere on or in the body and on other species.

plankton: small aquatic plants and animals, not necessarily microscopic.

polychaetes: group name for segmented marine worms that have swimming appendages with many chaetae or bristles.

purse seine: any net that is used to surround schools of fish and which is then closed or pursed at the bottom to trap the encircled fish.

sexual dimorphism: a phenomenon in which males and females differ markedly in shape, color, or other external features.

teleosts: fishes having ossified skeletons as opposed to elasmobranchs which have cartilaginous skeletons.

trammel net: a net having three walls of mesh, the innermost comprised of smaller mesh than those on either side; especially useful for catching halibut and a few other species.

viviparous: producing living young instead of eggs.

INDEX OF SCIENTIFIC NAMES

[177]

Index of Common Names